1 Introduction

Computer users have become accustomed to an exponential increase in computing speed and capacity over the past few decades. Gordon Moore observed in 1965 that chip capacity doubled every year. Although the growth rate has slowed to "only" doubling about every 18 months, the geometric increase predicted by "Moore's law", as it is called, has held for over 40 years. Today's high-end PCs have the same power as machines that were considered supercomputers not long ago. Software advances have been equally dramatic, perhaps most familiar to the average user in the form of computer graphics. The crude colored dots and flat polygons in computer games of 20 years ago have been replaced by the near-photorealistic graphics of today's video games and movies.

An enormous amount of computing power is required for the complex software used in computer animations, molecular biology analyses, computational fluid dynamics, global climate and economic modeling, worldwide credit card processing, and a host of other sophisticated applications. The demands of these problem domains have led researchers to develop distributed computing systems harnessing the power of thousands, and in

---

[1] Preprint of article to appear in *Advances in Computers*, Marvin Zelkowitz, ed.
Available at http://hissa.nist.gov/~black/Papers/quantumCom.html

some cases more than a million, processors into clusters. Yet there are limits to this approach. Adding more processors increases the computing capacity of these clusters only linearly, yet many problems, particularly in physics and computer science, increase exponentially with the size of their inputs. The computing demands of these problems seem to be inherent in the problems themselves; that is, the overwhelming consensus is that no possible algorithm executable on a Turing machine, the standard model of computing, can solve the problem with less than exponential resources in time, memory, and processors.

The doubling of computing power every 18 months has enabled scientists to tackle much larger problems than in the past, but even Moore's law has limits. For each new chip generation, the doubling of capacity means that about half as many atoms are being used per bit of information. But when projected into the future, this trend reaches a limit of one atom per bit of information sometime between 2010 and 2020. Does this mean that improvements in computing will slow down at that point? Fortunately, the answer is "not necessarily." One new technology, quantum computing, has the potential to not only continue, but in fact dramatically *increase* the rate of advances in computing power, at least for some problems. The key feature of a quantum computer is that it can go beyond the Turing machine model of computation. That is, *there are functions that can be computed on a quantum computer that cannot be effectively computed with a conventional computer (i.e., a classical Turing machine.)* This remarkable fact underlies the enormous power of quantum computing.

Why all the excitement now? In 1982 Richard Feynman pointed out [Feyn82] that simulating some quantum mechanical systems took a huge amount of classical resources. He also suggested the complement, that if the quantum effects could be harnessed, they may be able to do a huge amount of classical computation. However, nobody had any idea how that might be done.

At about the same time, David Deutsch tried to create the most powerful model of computation consistent the laws of physics. In 1985 he developed the notion of a Universal Quantum Computer based on the laws of quantum mechanics. He also gave a simple example suggesting that quantum computers may be more powerful than classical computers. Many people improved on this work in the following decade. The next breakthrough came in 1994 when Peter Shor demonstrated [Shor95] that quantum computers could factor large numbers efficiently. This was especially exciting since it is widely believed that no efficient factoring algorithm is possible for classical computers. One limitation still dimmed the lure of quantum computing.

Quantum effects are exceeding fragile. Even at atomic sizes, noise tends to quickly distort quantum behavior and squelch non-classical phenomenon. How could a quantum computer undergo the hundreds or thousands of processing steps needed for even a single algorithm without some way to compensate for errors? Classical computers use millions and even billions of atoms or electrons to smooth out random noise. Communication, storage, and processing measures and compares bits along the way to detect and correct small errors before they accumulate to cause incorrect results, distorted messages, or even system crashes. But measuring a quantum mechanical system causes the quantum system to change.

An important breakthrough came in 1996 when Andrew Steane and independently, Richard Calderbank and Peter Shor, discovered methods of encoding quantum bits, or "qubits", and measuring group properties so that small errors can be corrected. These ingenious methods use collective measurement to identify characteristics of a group of qubits, for example, parity. Thus it is conceivable to compensate for an error in a single qubit while preserving the information encoded in the collective quantum state.

Although a lot of research and engineering remain, today we see no theoretical obstacles to quantum computation and quantum communication. In this article, we review quantum computing and communications, current status, algorithms, and problems that remain to be solved. Section 2 gives the reader a narrative tutorial on quantum effects and major theorems of quantum mechanics. Section 3 presents the "Dirac" or "ket" notation for quantum mechanics and mathematically restates many of the examples and results of the preceding section. Section 4 goes into more of the details of how a quantum computer might be built and explains some quantum computing algorithms, such as Shor's for factoring, Deutsch's for function characterization, and Grover's for searching, and error correcting schemes. Section 5 treats quantum communication and cryptography. We end with an overview of physical implementations in Section 6.

2   The Surprising Quantum World

Subatomic particles act very differently from objects in the everyday world. Particles can have a presence in several places at once. Also two well-separated particles may have intertwined fates, and the observation of one of the particles will cause this remarkable behavior to vanish. Quantum mechanics describes these, and other physical phenomena extraordinarily well.

We begin with a simple experiment that you can do for a few dollars worth of equipment. Begin with a beam of light passing through a polarizer, as in Figure 1.

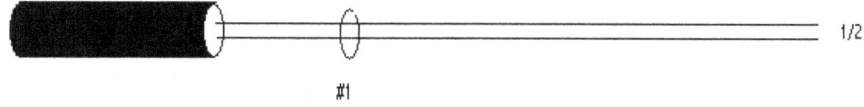

Figure 1. Polarizer dims beam by half.

A typical beam, such as from the sun or a flashlight, has its intensity reduced by half. Suppose we add another polarizer after the first. As we rotate the polarizer, the beam brightens and dims until it is gone[2], as depicted in Figure 2.

---

[2] Real polarizers are not perfect, of course, so a little light always passes.

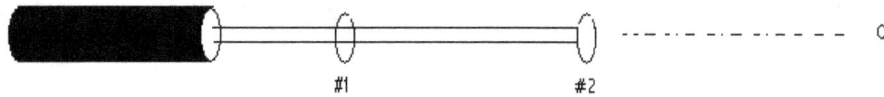

Figure 2. Two orthogonal polarizers extinguish the beam.

Leaving the two polarizers at the minimum, add a third polarizer between them, as shown in Figure 3. As we rotate it, we can get some light to pass through! How can adding another filter *increase* the light getting through?

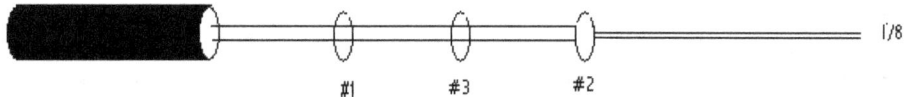

Figure 3. A third polarizer can partially restore the beam!

Although it takes extensive and elaborate experiments to prove that the following explanation is accurate, we assure you it is. Classical interpretations of these results are misleading at best.

To begin the explanation, photons have a characteristic called "polarization". After passing through polarizer #1, all the photons of the light beam are polarized in the same direction as the polarizer. If a polarizer is set at right angles to polarizer #1, the chance of a photon getting through both polarizers is 0, that is, no light gets through. However, when the polarizer in the middle is diagonal to polarizer #1, half the photons pass through the first two polarizers. More importantly, the photons are now oriented diagonally. Half the diagonally oriented photons can now pass through the final polarizer. Because of their relative orientations, each polarizer lets half the photons through, so a total of 1/8 passes through all three polarizers.

2.1 Sidebar: Doing the Polarization Experiment Yourself

You can do the polarization experiment at home with commonly available materials costing a few dollars. You need a bright beam of light. This could be sunlight shining through a hole, a flashlight, or a laser pointer.

For polarizers you can use the lens from a pair of polarizing sunglasses. You can tell if sunglasses are polarizing by holding two pair, one behind the other, and looking through the left (or right) lens in series. Rotate one pair of sunglasses relative to the other while keeping the lens in line. If the scene viewed through the lens darkens and lightens as one is rotated, they are polarizing. You can also buy gray light polarizing glasses or plastic sheets on the World Wide Web.

Carefully free the lens. One lens can be rigidly attached to a support, but the others must be able to rotate. Shine the light through polarizer #1. Put polarizer #2 in the beam well after polarizer #1. Rotate 2 until the least amount of light comes through. Now put polarizer 3 between 1 and 2. Rotate it until the final beam of light is its brightest. By trying different combinations of lenses and rotations, you can verify that the lenses are at 45° and 90° angles from each other.

2.2 Returning to the Subject at Hand ...

After we develop the mathematics, we will return to this example in Sect. 3.3 and show how the results can be derived. The mathematical tools we use are quantum mechanics. Quantum mechanics describes the interactions of electrons, photons, neutrons, etc. at atomic and subatomic scales. It does not explain general relativity, however. Quantum mechanics makes predictions on the atomic and subatomic scale that are found to be extremely accurate and precise. Experiments support this theory to better accuracy than any other physical theory in the history of Science.

The effects we see at the quantum level are very different from those we see in the every-day world. So, it should not come as a surprise that different mathematics is used. This section presents fundamental quantum effects and describes some useful laws that follow from those.

2.3 The Four Postulates of Quantum Mechanics

Quantum Mechanics is mathematically very well defined and is a framework for defining physical systems. This powerful framework defines what may and may not happen in quantum mechanical systems. Quantum Mechanics itself does not give the details of any one particular physical system.

Some analogies may help. Algebraic groups have well defined properties, such as that operations are closed. Yet, the definition of a group does not give the group of rotations in 3-space or addition on the integers. Likewise, the rules for a role-playing game limit what is and is not allowed, but don't describe individuals or scenarios.

Quantum mechanics consists of four postulates. [NC00 pp 80-94]

Postulate 1 *Any isolated quantum system can be completely mathematically characterized by a state vector in a Hilbert space.*

A Hilbert space is a complex vector space with inner product. Experiments show there is no need for other descriptions since all the interactions, such as momentum transfer, electric fields, and spin conservation, can be included within the framework. The postulates of quantum mechanics, by themselves, do not tell us what the appropriate Hilbert space is for a particular system. Rather, physicists work long and hard to determine the best approximate model for their system. Given this model, their experimental results can be described by a vector in this appropriate Hilbert space.

The notation we will explain in Sect. 3 cannot express all possible situations, such as if we wish to track our incomplete knowledge of a physical system, but suffices for this paper. There are more elaborate mathematical schemes that can represent as much quantum information as we need.

Postulate 2 *The time evolution of an isolated quantum system is described by a unitary transformation.*

Physicists use the term "time evolution" to express that the state of a system is changing solely due to passage of time, for instance, particles are moving or interacting. If the quantum system is completely isolated from losses to the environment or influences from outside the system, any evolution can be captured by a unitary matrix expressing a transformation on the state vector.

Again, pure quantum mechanics doesn't tell us what the transformation is, but provides the framework into which experimental results must fit. The corollary is that isolated quantum systems are reversible.

Postulate 3 *Only certain sets of measurements can be done at any one time. Measuring projects the state vector of the system onto a new state vector.*

This is the so-called collapse of the system. From a mathematical description of the set of measurements, one can determine the probability of a state yielding each of the measurement outcomes.

One powerful result is that arbitrary quantum states cannot be measured with arbitrary accuracy. No matter how delicately done, the very first measurement forever alters the state of the system. We discuss this in more detail in Section 2.6.

The measurements in a set, called "basis", are a description of what can be observed. Often quantum systems can be described with many different, but related, bases. Analogously, positions in the geometric plane may be given as pairs of distances from the origin along orthogonal, or perpendicular, axes, such as X and Y. However, positions may also be given as pairs of distances along the diagonal lines X=Y and X= -Y, which form an equally valid set of orthogonal axes. A simple rotation transforms between coordinates in either basis. Polar coordinates provide yet another alternative set of coordinates. Although it may be easier to work with one basis or another, it is misleading to think that coordinates in one basis are *the* coordinates of a position, to the exclusion of others.

Postulate 4 *The state space of a composite system is the tensor products of the state spaces of the constituent systems.*

Herein lies a remarkable opportunity for quantum computing. In the everyday world, the state space of a system composed of several subsystems is the product of the state spaces of the subsystems that constitute it. However, quantum mechanical systems can become very complicated very fast. The negative view is to realize how much classical computation we need to simulate even simple systems of, say, 10 particles. The positive view is to wonder if this enormously rich state space might be harnessed for very powerful computations.

2.4 Superposition

As can be seen from the polarization experiment above, very tiny entities may behave very differently from macroscopic bodies. An everyday solid object has a definite position, velocity, etc. But at the quantum scale, particle characteristics are best described as blends or superpositions of base values. When measured, we get a definite value. However, between measurement events, any consistent mathematical model must allow for the potential, or "amplitude", of several states at once. Another example may provide a more intuitive grasp.

2.4.1 Young's Double Slit Experiment

In 1801, using only a candle for a light source, Thomas Young performed an experiment whose results can only be explained if light acts as a wave [You07]. Young shined the light through two parallel slits onto a surface, as diagrammed in Figure 4, and saw a pattern of light and dark bands. The wavy line on the right graphs the result; light intensity is the horizontal axis, increasing to the right. This is the well-known interference effect: waves, which cancel and reinforce each other, produce this pattern.

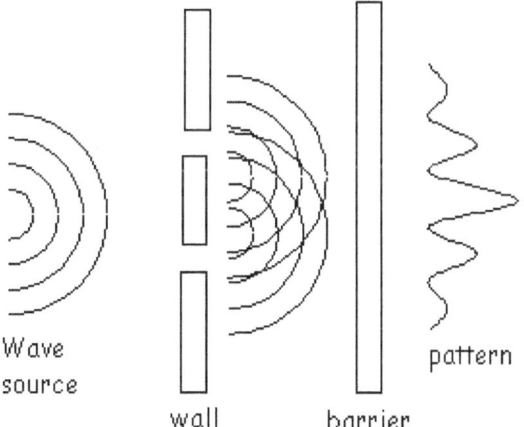

Figure 4. Young's Double Slit Experiment

Imagine, in contrast, a paintball gun pointing at a wall in which two holes have been drilled, beyond which is a barrier, as shown in Figure 5. The holes are just big enough for a single paintball to get through, though the balls may ricochet from the sides of the holes. The gun shoots at random angles, so only a few of the paintballs get through. If one of the holes is covered up, the balls that get through will leave marks on the wall, with most of the marks concentrated opposite the hole and others scattered in a bell curve

(P1) to either side of the hole, as shown in the figure. If only the second hole is open, a similar pattern (P2) emerges on the barrier immediately beyond the hole. If both holes are open, the patterns simply add. The paint spots are especially dense where the two patterns overlap, resulting in a bimodal distribution curve that combines P1 and P2. No interference is evident.

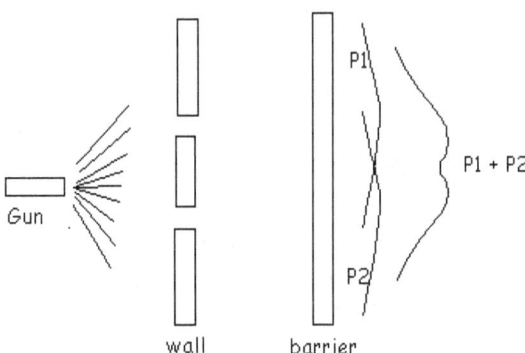

Figure 5. Paintballs fired at a wall

What happens when electrons are fired at two small slits, as in Figure 6? Surprisingly, they produce the same wave pattern of Figure 4. That is, the probability of an electron hitting the barrier at a certain location varies in a pattern of alternating high and low, rather than a simple bimodal distribution. This occurs even when electrons are fired one at a time. Similar experiments have been done with atoms and even large molecules of carbon-60 ("buckeyballs"), all demonstrating wave-like behavior of matter. So something "wave-like" must be happening at small scales.

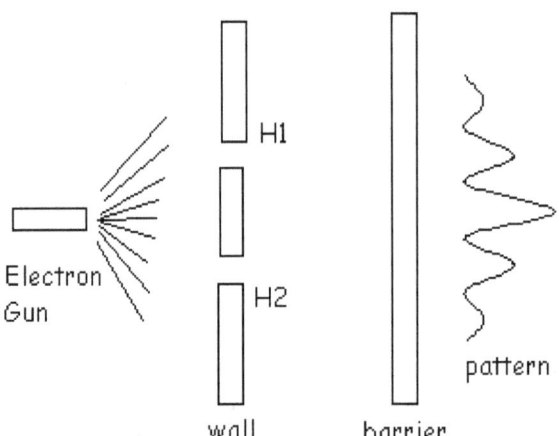

Figure 6. Double slit experiment with electrons

2.4.2 Explaining the Double Slit Experiment

How do we explain these results? If a wave passes through the slits, we can expect interference, canceling or reinforcing, resulting in a pattern of light and dark lines. But how can individual electrons, atoms, or molecules, fired one at a time, create interference patterns? A desperate classical explanation might be that the particles split, with one part passing through each hole. But this is not the case: if detectors are placed at H1 or at H2 or in front of the barrier, only one particle is ever registered at a time. (Remarkably, if a detector is placed at H1 or H2, the pattern follows Figure 5. More about this effect later.)

The quantum mechanical explanation is that particles may be in a "superposition" of locations. That is, an electron is in a combination of state "at H1" and "at H2". An everyday solid object has a definite position, mass, electric charge, velocity, etc. But at the quantum scale, particle characteristics are best described as blends or superpositions of base values. When measured, we always get a definite value. However, between measurement events, any consistent mathematical model must potentially allow for an arbitrary superposition of many states. This behavior is contrary to everyday experience, of course, but thousands of experiments have verified this fact: *a particle can be in a superposition of several states at the same time.* When measured, the superposition collapses into a single state, losing any information about the state before measurement.

The photons in the beam-and-filters experiment are in a superposition of polarizations. When polarizer #1 tests the photon for vertical or horizontal polarization, the photon either emerges polarized vertically, or it doesn't emerge. No information about prior states is maintained. It is not possible to determine whether it had been vertical, diagonal, or somewhere in between. Since vertical polarization is a superposition, or combination, of diagonal polarizations, some of the vertically polarized photons pass through the middle polarizer and emerge polarized diagonally. Half of the now-diagonally polarized photons will pass through the final, horizontal polarizer.

2.5 Randomness

In the beam-and-filters experiment, randomly some photons emerge polarized while others do not emerge at all. This unpredictability is not a lack of knowledge. It is not that we are missing some full understanding of the state of the photons. The random behavior is truly a part of nature. We cannot, even in principle, predict which of the photons will emerge and which will not.

This intrinsic randomness may be exploited to generate cryptographic keys or events that are not predictable. But it also means that the unpredictability of some measurements is not merely an annoying anomaly to be reduced by better equipment, but an inherent property in quantum computation and information. Even though an individual measurement may be arbitrary, the statistical properties are well defined. Therefore we may take advantage of the randomness or unpredictability in individual outcomes.

We can make larger or more energetic systems that are more predictable, but then the quantum properties, which may be so useful, disappear, too.

2.6 Measurement

As opposed to being an objective, external activity, in quantum mechanics measuring a system is a significant step. A measurement is always with regard to two or more base values. In the photon polarization experiment, the bases are orthogonal directions: vertical and horizontal, two diagonals, 15° and 105°, etc. The basis for other systems may be in terms of momentum, position, energy level, or other physical quantities.

When a quantum system is measured, it collapses into one of the measurement bases. No information about previous superpositions remains. We cannot predict into which of the bases a system will collapse, however given a known state of the system, we can predict the probability of measuring each basis.

2.7 Entanglement

Even more surprising than superposition, quantum theory predicts that entities may have correlated fates. That is, the result of a measurement on one photon or atom leads instantaneously to a correlated result when an entangled photon or atom is measured.

For a more intuitive grasp of what we mean by "correlated results", imagine that two coins could be entangled (there is no known way of doing this with coins of course). Imagine one is tossing a coin. Careful records show it comes up "heads" about half the time and "tails" half the time, but any one result is unpredictable. Tossing another coin has similar, random results. But surprisingly, the records of the coin tosses show a correlation! When one coin comes up heads, the other coin comes up tails and vice versa. We say that the state of the two coins is entangled. Before the measurement (the toss), the outcome is unknown, but we know the outcomes will be correlated. As soon as either coin is tossed (measured), the fate of tossing the other coin is sealed. We cannot predict in advance what an individual coin will do, but their results will be correlated: once one is tossed, there is no uncertainty about the other.

This imaginary coin tossing is only to give the reader a sense of entanglement. Although one might come up with a classical explanation for these results, multitudes of ingenious experiments have confirmed the existence of entanglement and ruled out any possible classical explanation. Over several decades, physicists have continually refined these experiments to remove loopholes in measurement accuracy or subtle assumptions. All have confirmed the predictions of quantum mechanics.

With actual particles any measurement collapses uncertainty in the state. A real experiment would manufacture entangled particles, say by bringing particles together and entangling them or by creating them with entangled properties. For instance, we can "downconvert" one higher energy photon into two lower energy photons which leave in directions which not entirely predictable. Careful experiments show that the directions are actually a superposition, not merely a random, unknown direction. However, since the momentum of the higher energy photon is conserved, the directions of the two lower energy photons are entangled. Measuring one causes both photons to collapse into one of the measurement bases. But once entangled, the photons can be separated by any distance, at any two points in the universe, yet measuring one will result in a perfectly correlated measurement for the other.

Even though measurement brings about a synchronous collapse regardless of the separation, entanglement doesn't let us transmit information. We cannot force the result of a measurement any more than we can force the outcome of tossing a fair coin (without interference).

2.8 Reversibility

Postulate 2 of quantum mechanics says that the evolution of an isolated system is reversible. In other words, any condition leading to an action also may bring about the reverse action in time-reversed circumstances. If we watch a movie of a frictionless pendulum, we cannot tell if the movie is being shown backwards or not. In either case, the pendulum moves according to the laws of momentum and gravity. If a beam of photons is likely to move an electron from a lower to a higher energy state, the beam is also likely to move an electron from the higher energy state to the lower one. (In fact, this is the "stimulated emission" of a laser.) This invertible procession of events is referred to as "unitary evolution." To preserve superposition and entanglement, we must use unitary evolutions.

An important consequence is that operations should be reversible. Any operation that loses information or is not reversible cannot be unitary, and may loose superposition and entanglement. Thus to guarantee that a quantum computation step preserves superposition and entanglement, it must be reversible.

Finding the conjunction of A AND B is not reversible: if the result is false, we do not know if A was false, or B was false, or both A and B were false. Thus a standard conjunction destroys superpositions and entanglements. However, suppose we set another bit, C, previously set to false, to the conjunction of A AND B, and keep the values of both A and B. This computation is reversible. Given any resulting state of A, B, and C, we can determine the state before the computation. Likewise all standard computations can be done reversibly, albeit with some extra bits. We revisit reversible computations in Sect. 4.1.1.

2.9 The Exact Measurement "Theorem"

Although quantum mechanics seems strange, it is a very consistent theory. Seemingly reasonable operations are actually inconsistent with the theory as a whole. For instance, one might wish to harness entanglement for faster-than-light or even instantaneous communication. Unfortunately, any measurement or observation collapses the state. Also unfortunately, it is impossible to tell with local information whether the observation preceded or followed the collapse: the observation gives the same random result in either case. Communicating with the person holding the other entangled particle, to determine some correlation, can only be done classically, that is, no faster than the speed of light. So entanglement cannot be used to transmit information faster than light and violate relativity.

If we could exactly measure the entire quantum state of a particle, we could determine if it were in a superposition. Alice and Bob could begin with two pairs of particles; let us call them the "T" pair, T1 and T2, and the "F" pair, F1 and F2. They manipulate them so T1 and T2 are entangled with each other and F1 and F2 are entangled with each other. Bob then takes T1 and F1 far away from Alice. If exact measurement were possible, Bob could continuously measure his particle T1 to see if it has collapsed into a definite state. To instantly communicate a "1", Alice observes her member of the "T" pair, T2, causing it to collapse. Because the "T" pair was entangled, Bob's particle, T1, simultaneously collapses into a definite state. Bob detects the collapse of T1, and writes down a "1." Similarly, a "0" bit could be transmitted instantly using the "F" pair if, indeed, exact measurement were possible. In fact, if we were able to exactly measure an unknown quantum state, it would lead to many inconsistencies.

2.10 The No-Cloning Theorem

One might be tempted to evade the impossibility of exact measurement by making many exact copies of particles and measuring the copies. If we could somehow manage to have an unlimited supply of exact copies, we could measure them and experimentally build up an exact picture of the quantum state of the original particle. However, the "No-Cloning Theorem" proves we cannot make an exact copy of an unknown quantum state. In Sect. 3.6 we prove a slightly simplified version of the theorem.

What about setting up an apparatus, say with polarizers, laser beams, magnetic fields, etc. which produces an unlimited number of particles, all in the same quantum state? We could make unlimited measurements in various bases, and measure the state to arbitrary accuracy. Indeed, this is what experimental physicists do. But it is a measurement of the result of a process, not the measurement of a single, unknown state.

Alternatively, if we could exactly measure an unknown quantum state, we could prepare as many particles as we wished in that state, effectively cloning. So the lack of exact measurement foils this attempt to clone, and the lack of cloning closes this route to measurement, maintaining the consistency of quantum mechanics.

3 The Mathematics of Quantum Mechanics

The ugly truth is that general relativity and quantum mechanics are not consistent. That is, our current formulations of general relativity and quantum mechanics give different predictions for extreme cases. We assume there is a "Theory of Everything" that reconciles the two, but it is still very much an area of thought and research. Since relativity is not needed in quantum computing, we ignore this problem. Let us emphasize that thousands of experiments that have been done throughout the world in the last 100 years are consistent with quantum mechanics.

We present a succinct notation and mathematics commonly used to formally express the notions of quantum mechanics. Although this formalization cannot express all the nuances, it is enough for this introductory article. More complete notations are given in various books on quantum mechanics.

3.1 Dirac or Ket Notation

We can represent the state of quantum systems in "Dirac" or "ket"[3] notation. ("Ket" rhymes with "let.") A qubit is a quantum system with two discrete states. These two states can be expressed in ket notation as $|0\rangle$ and $|1\rangle$. An arbitrary quantum state is often written $|\Psi\rangle$. State designations can be arbitrary symbols. For instance, we can refer to the polarization experiment in Section 2 using the bases $|\uparrow\rangle$ and $|\rightarrow\rangle$ for vertical and horizontal polarization and $|\nearrow\rangle$ and $|\nwarrow\rangle$ for the two orthogonal diagonal polarizations. (Caution: although we use an up-arrow for vertical, "up" and "down" polarization are the same thing: they are both vertical polarization. Likewise be careful not to misinterpret the right or diagonal arrows.)

A quantum system consisting of two or more quantum states is the tensor product of the separate states in some fixed order. Suppose we have two photons, P1 and P2, where P1 has the state $|P1\rangle$, and P2 has the state $|P2\rangle$. We can express the state of the joint system as $|P1\rangle \otimes |P2\rangle$, or we can express it as $|P2\rangle \otimes |P1\rangle$. The particular order doesn't matter as long as it is used consistently.

For brevity, the tensor product operator is implicit between adjacent states. The above two photon system is often written $|P1\,P2\rangle$. Since the order is typically implicit, the ket is usually written without indices, thus: $|PP\rangle$. Ket "grouping" is associative, therefore a single ket may be written as multiple kets for clarity: $|0\rangle|0\rangle|0\rangle$, $|0\rangle|00\rangle$, and $|00\rangle|0\rangle$, all mean $|0_1 0_2 0_3\rangle$. Bases are written in the same notation using kets. For example, four orthogonal bases of a two qubit system are $|00\rangle$, $|01\rangle$, $|10\rangle$, and $|11\rangle$. Formally, a ket is just a column vector.

---

[3] The name comes from "bracket." P. A. M. Dirac developed a shorthand "bracket" notion to express the outer product of state vectors, $\langle\Psi|\Psi\rangle$. In most cases the column vector, or right-hand side, can be used alone. Being the second half of a bracket, it is called a ket.

3.2 Superpositions and Measurements

Superpositions are written as a sum of states, each with an "amplitude" which may be a complex number. For instance, if an electron has a greater probability of going through the top slit in Figure 6, its position might be $\sqrt{1/4}|H1\rangle + \sqrt{3/4}|H2\rangle$. The polarization of a photon that is in an equal superposition of vertical and horizontal polarizations may be written as $1/\sqrt{2}|\uparrow\rangle + 1/\sqrt{2}|\rightarrow\rangle$. In general, a two-qubit system is in the state $a|00\rangle + b|01\rangle + c|10\rangle + d|11\rangle$.

The norm squared of the amplitude of a state is the probabilities of measuring the system in that state. The general two qubit system from above will be measured in state $|00\rangle$ with probability $|a|^2$. Similarly, the system will be measured in states $|01\rangle$, $|10\rangle$, or $|11\rangle$ with probabilities, $|b|^2$, $|c|^2$, and $|d|^2$ respectively. Amplitudes must be used instead of probabilities to reflect quantum interference and other phenomena.

Because a measurement always finds a system in one of the basis states, the probabilities sum to one. (The requirement that they sum to one is a reflection of the basic conservation laws of physics.) Hence the sum of norm squared amplitudes must always sum to one, also. Amplitudes that nominally do not sum to one are understood to be multiplied by an appropriate scaling factor to "normalize" them so they *do* sum to one.

A measurement collapses the system into one of the bases of the measurement. The probability of measuring the system in, or equivalently, collapsing the system into any one particular basis is the norm squared of the probability. Hence, for the location distribution $\sqrt{1/4}|H1\rangle + \sqrt{3/4}|H2\rangle$, the probability of finding an electron at location H1 is $\left|\sqrt{1/4}\right|^2 = 1/4$, and the probability of finding an electron at H2 is $\left|\sqrt{3/4}\right|^2 = 3/4$. After measurement, the electron is either in the state $|H1\rangle$, that is, at H1, or in the state $|H2\rangle$, that is, at H2, and there is no hint that the electron ever had any probability of being anywhere else. If measurements are done at H1 or H2, the interference disappears resulting in the simple bimodal distribution shown in Figure 5.

3.3 The Polarization Experiment, Again

Just as geometric positions may be equally represented by different coordinate systems, quantum states may be expressed in different bases. A vertically polarized photon's state may be written as $|\uparrow\rangle$. It may just as well be written as a superposition of two diagonal bases $1/\sqrt{2}|\nearrow\rangle + 1/\sqrt{2}|\nwarrow\rangle$. Likewise a diagonally polarized photon $|\nwarrow\rangle$ may be viewed as being in a superposition of vertical and horizontal polarizations $1/\sqrt{2}|\uparrow\rangle + 1/\sqrt{2}|\rightarrow\rangle$. For the polarization is $|\nearrow\rangle$, the superposition is $1/\sqrt{2}|\uparrow\rangle - 1/\sqrt{2}|\rightarrow\rangle$; note the sign change. In both cases, the amplitudes squared, $\left|1/\sqrt{2}\right|^2$ and $\left|-1/\sqrt{2}\right|^2$, still sum to one. We can now express the polarization experiment formally.

The first polarizer "measures" in some basis, which we can call $|\uparrow\rangle$ and $|\rightarrow\rangle$. Regardless of previous polarization, the measurement leaves photons in either $|\uparrow\rangle$ or $|\rightarrow\rangle$, but only passes photons which are, say $|\uparrow\rangle$. If the incoming beam is randomly polarized, half the photons collapse into, or are measured as, $|\uparrow\rangle$ and passed, which agrees with the observation that the intensity is halved.

A second polarizer, tilted at an angle, θ, to the first, "measures" in a tilted basis $|\nearrow_{\cos\theta}\rangle$ and $|\nwarrow_{\sin\theta}\rangle$. Photons in state $|\uparrow\rangle$ can also be considered to be in the superposition $\cos\theta|\nearrow_{\cos\theta}\rangle + \sin\theta|\nwarrow_{\sin\theta}\rangle$. The second polarizer measures photons in the tilted basis, and passes only those collapsing into $|\nearrow_{\cos\theta}\rangle$. Since the chance[4] of a photon collapsing into that state is $\cos^2\theta$, the intensity of the resultant beam decreases to zero as the polarizer is rotated to 90°. With polarizer #2 set at right angles, it measures with the same basis as polarizer #1, that is $|\uparrow\rangle$ and $|\rightarrow\rangle$, but only passes photons with state $|\rightarrow\rangle$.

When polarizer #3 is inserted, it is rotated to a 45° angle. The vertically polarized, that is $|\uparrow\rangle$, photons from polarizer #1 can be considered to be in the superposition

$\cos 45°|\nearrow_{\cos 45°}\rangle + \sin 45°|\nearrow_{\sin 45°}\rangle = 1/\sqrt{2}|\nearrow\rangle + 1/\sqrt{2}|\nwarrow\rangle$. So they have a $\left|1/\sqrt{2}\right|^2 = 1/2$ chance of collapsing into state $|\nearrow\rangle$ and being passed. These photons encounter polarizer #2, where they can be considered to be in the superposition $\cos 45°|\uparrow_{\cos 45°}\rangle + \sin 45°|\rightarrow_{\sin 45°}\rangle = 1/\sqrt{2}|\uparrow\rangle + 1/\sqrt{2}|\rightarrow\rangle$. So they again have a $\left|1/\sqrt{2}\right|^2 = 1/2$ chance of collapsing, now into state $|\rightarrow\rangle$, and being passed. Thus the chance of an arbitrary photon passing through all three polarizers is $1/2 \times 1/2 \times 1/2 = 1/8$ agreeing with our observation.

3.4 Expressing Entanglement

In the Dirac or ket notation, the tensor product, $\otimes$, distributes over addition, e.g., $|0\rangle \otimes \left(1/\sqrt{2}|0\rangle + 1/\sqrt{2}|1\rangle\right) = \left(1/\sqrt{2}|00\rangle + 1/\sqrt{2}|01\rangle\right)$. Another example is that the tensor product of equal superpositions is an equal superposition of the entire system:

$$\left(\frac{1}{\sqrt{2}}|0\rangle + \frac{1}{\sqrt{2}}|1\rangle\right) \otimes \left(\frac{1}{\sqrt{2}}|0\rangle + \frac{1}{\sqrt{2}}|1\rangle\right)$$

---

[4] To double check consistency, note that the probability of seeing either state is $\cos^2\theta + \sin^2\theta = 1$.

$$= \left(\frac{1}{\sqrt{2}}|0\rangle \frac{1}{\sqrt{2}}|0\rangle + \frac{1}{\sqrt{2}}|0\rangle \frac{1}{\sqrt{2}}|1\rangle + \frac{1}{\sqrt{2}}|1\rangle \frac{1}{\sqrt{2}}|0\rangle + \frac{1}{\sqrt{2}}|1\rangle \frac{1}{\sqrt{2}}|1\rangle\right)$$

$$= \frac{1}{2}\bigl(|00\rangle + |01\rangle + |10\rangle + |11\rangle\bigr)$$

Note that the square of each amplitude gives a 1/4 chance of each outcome, which is what we expect.

If a state cannot be factored into products of simpler states, it is "entangled." For instance, neither $1/2|00\rangle + \sqrt{3/4}|11\rangle$ nor $1/\sqrt{2}\bigl(|Heads\ Tails\rangle + |Tails\ Heads\rangle\bigr)$ can be factored into a product of states. The latter state expresses the entangled coin tossing we discussed in Sect. 2.7. When we toss the coins (do a measurement), we have equal chances of getting $|Heads\ Tails\rangle$ (heads on the first coin and tails on the second) or $|Tails\ Heads\rangle$ (tails on the first coin and heads on the second). If we observe the coins separately, they appear to be completely classical, fair coins: heads or tails appear randomly. But the records of the two coins are correlated: when one comes up heads, the other comes up tails and vice versa.

3.5 Unitary Transforms

Postulate 2 states that all transformations of an isolated quantum system are unitary. In particular, they are linear. If a system undergoes decoherence or collapse because of some outside influence, the transformation is not necessarily unitary. But when an entire system is considered in isolation from any other influence, all transformations are unitary.

3.6 Proof of No-Cloning Theorem

With postulate 2, we can prove a slightly simplified version of the No-Cloning Theorem. (A comprehensive version allows for arbitrary ancillary or "work" qubits.) We begin by formalizing the theorem. We hypothesize that there is some operation, U, which exactly copies an arbitrary quantum state, Ψ, onto another particle. Its operation would be written as follows.

$$U|\Psi\rangle|0\rangle = |\Psi\rangle|\Psi\rangle$$

Does this hypothetical operator have a consistent definition for a state that is a superposition? In Dirac notation, what is the value of $U(a|0\rangle + b|1\rangle)|0\rangle$?

Recall that tensor product distributes over superposition. One derivation is to distribute the tensor product first, then distribute the clone operation, and finally perform the hypothetical clone operation.

$$\begin{aligned} U(a|0\rangle + b|1\rangle)|0\rangle &= U(a|0\rangle|0\rangle + b|1\rangle|0\rangle) \\ &= Ua|0\rangle|0\rangle + Ub|1\rangle|0\rangle \\ &= a|0\rangle a|0\rangle + b|1\rangle b|1\rangle \\ &= a^2|00\rangle + b^2|11\rangle \end{aligned}$$

But if we evaluate the clone operation first then distribute, we get the following.

$$\begin{aligned} U(a|0\rangle + b|1\rangle)|0\rangle &= (a|0\rangle + b|1\rangle)(a|0\rangle + b|1\rangle) \\ &= a|0\rangle a|0\rangle + a|0\rangle b|1\rangle + b|1\rangle a|0\rangle + b|1\rangle b|1\rangle \\ &= a^2|00\rangle + ab|01\rangle + ab|10\rangle + b^2|11\rangle \end{aligned}$$

The derivations are different! The mathematics should be consistent unless we're trying something impossible, like dividing by zero. Since the only questionable step was assuming the existence of a cloning operation, we conclude that a general cloning operation is inconsistent with the laws of quantum mechanics.

Notice that if a is zero or b is zero, the two derivations *do* give the same result. But a and b are amplitudes (like probabilities) of states in the superposition. If one or the other is zero, there was actually no superposition to begin with, and this proof doesn't apply. In fact, in the absence of arbitrary superposition, we can clone.

If we know that a particle is either in state $|0\rangle$ or in state $|1\rangle$, we can simply measure the particle. We then set any number of other particles to that same state, effectively copying the state of the particle. In this case we know something about the original state of the particle. So this "loophole" does not invalidate the theorem that we cannot clone a completely unknown state.

In Sect. 5.6 we explain how we *can* move, or "teleport", an unknown state to a distant particle. But the state on the original particle is destroyed in the process. So we still end up with just one instance of a completely unknown state.

4 Quantum Computing

We have seen that phenomena and effects at quantum scales can be quite different from those we are used to. The richness of these effects tantalize us with the possibility of far faster computing, when we manage to harness these effects. But how can we turn these effects in gates and computers? How fast might they solve problems? Are these merely theoretical ideals, like a frictionless surface or noiseless measurement, or is there hope of building an actual device? This section discusses how quantum effects can be harnessed to create gates, assesses the potential for quantum algorithms, and outlines ways of dealing with imperfect operations and devices.

4.1 Quantum Gates and Quantum Computers

Digital computers, from microprocessors to supercomputers, from the tiny chips running your wristwatch or microwave to continent-spanning distributed systems that handle worldwide credit card transactions, are built of thousands or millions of simple gates. Each gate does a single logical operation; such as producing a 1 if all its inputs are 1 (otherwise producing a 0 if any input is 0) or inverting a 1 to a 0 and a 0 to a 1. From these simple gates, engineers build more complex circuits that add or multiply two numbers, select a location in memory, or choose which instructions to do next depending on the result of an operation. From these circuits, engineers create more and more complex modules until we have computers, CD players, aircraft navigation systems, laser printers, and cell phones. Although computer engineers still must deal with significant concerns, such as transmitting signals at gigahertz rates, getting a million gates to function in exact lockstep, or storing ten billion bits without losing a single one, conceptually once we can build simple gates, the rest is "merely" design.

Quantum computing appears to be similar: we know how to use quantum effects to create quantum gates or operations, we have ideas about combining gates into meaningful modules, and we have an increasing body of work about how to do quantum computations reliably, even with imperfect components. Researchers are optimistic because more work brings advances in both theory and practice.

In classical computer design, one basic gate is the AND gate. However, as we described in Sect. 2.8, an AND gate is not reversible. A basic, reversible quantum gate is the "controlled-not" or CNOT gate. It is represented as in Fig. 7. The two horizontal lines, labeled $|\varphi\rangle$ and $|\psi\rangle$, represent two qubits. The gate is the vertical line with connections to the qubits.

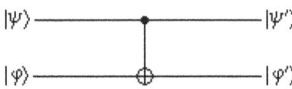

Figure 7. A CNOT gate

The top qubit, labeled $|\psi\rangle$ and connected with the dot, is the gate's control. The bottom qubit, labeled $|\varphi\rangle$ and connected with ⊕, is the "data". The data qubit is inverted if the control qubit is 1. If the control is 0, the data qubit is unchanged. Table 1 shows the operation of CNOT. Typically we consider the inputs to be on the left (the $|\varphi\rangle$ and $|\psi\rangle$), and the outputs to be on the right. Since CNOT is reversible, it is not unreasonable to consider the right hand side (the $|\varphi'\rangle$ and $|\psi'\rangle$) the "inputs" and the left hand side the "outputs"! That is, we can run the gate "backwards". The function is still completely determined: every possible "input" produces exactly one "output".

| $|\psi\rangle$ | $|\varphi\rangle$ | $|\psi'\rangle$ | $|\varphi'\rangle$ |

$\lvert 0\rangle$	$\lvert 0\rangle$	$\lvert 0\rangle$	$\lvert 0\rangle$
$\lvert 0\rangle$	$\lvert 1\rangle$	$\lvert 0\rangle$	$\lvert 1\rangle$
$\lvert 1\rangle$	$\lvert 0\rangle$	$\lvert 1\rangle$	$\lvert 1\rangle$
$\lvert 1\rangle$	$\lvert 1\rangle$	$\lvert 1\rangle$	$\lvert 0\rangle$

Table 1. Function of the CNOT gate

So far, this is just the classical exclusive-OR gate. What happens when the control is a superposition? The resultant qubits are entangled. In the following, we apply a CNOT to the control qubit, an equal superposition of $\lvert 0_\psi\rangle$ and $\lvert 1_\psi\rangle$ (we use the subscript ψ to distinguish the control qubit), and the data qubit, $\lvert 0\rangle$.

$$\mathrm{CNOT}\bigl(1/\sqrt{2}\,(\lvert 0_\psi\rangle + \lvert 1_\psi\rangle)\otimes \lvert 0\rangle\bigr) = 1/\sqrt{2}\,\bigl(\mathrm{CNOT}\lvert 0_\psi 0\rangle + \mathrm{CNOT}\lvert 1_\psi 0\rangle\bigr)$$

$$= 1/\sqrt{2}\,\bigl(\lvert 0_\psi 0\rangle + \lvert 1_\psi 1\rangle\bigr)$$

What does this mean? One way to understand it is to measure the control qubit. If the result of the measurement is 0, the state has collapsed to $\lvert 0_\psi 0\rangle$, so we will find the data qubit to be 0. If we measure a 1, the state collapsed to $\lvert 1_\psi 1\rangle$, and the data qubit is 1. We could measure the data qubit first and get much the same result. These results are consistent with Table 1.

So how might we build a CNOT gate? We review several possible implementations in Sect. 6, but sketch one here. Suppose we use the state of the outermost electron of a sodium atom as a qubit. An electron in the ground state is a logical 0, and an excited electron is a logical 1. An appropriate pulse of energy will flip the state of the qubit. That is, it will excite an outer electron in the ground state, and "discharge" an excited electron. To make a CNOT gate, we arrange a coupling between two atoms such that if the outer electron of the control atom is excited, the outer electron of the data atom flips when we apply a pulse. If the control atom is not excited, the pulse has no effect on the data atom.

As can be guessed from this description, the notion of wires and gates, as represented schematically in Fig. 7, might not be used in an actual quantum computer. Instead, different tuned and selected minute energy pulses may cause qubits to interact and change their states.

A more easily used quantum gate is the controlled-controlled-not or C2NOT gate. It has two control qubits and one data qubit, as represented schematically in Fig. 8. It is similar to the CNOT: the data qubit is inverted if both the control qubits are 1. If either is 0, the data qubit is unchanged. We can easily make a reversible version of the classical AND gate. To find A AND B, use A and B as the controls and use a constant 0 as the data. If A and B are both 1, the 0 is flipped to a 1. Otherwise it remains 0.

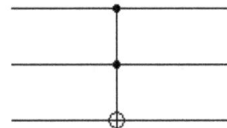

Figure 8. A C2NOT gate

Many other basic quantum gates have been proposed [NC00, Chapter 4]. Using these gates as building blocks, useful functions and entire modules have been designed. In short, we can conceptually design complete quantum computing systems. In practice there are still enormous, and perhaps insurmountable, engineering tasks before realistic quantum computing is available. For instance, energy pulses are never perfect, electrons don't always flip when they are supposed to, and stray energy may corrupt data. Sect. 4.3 explains a possible approach to handling such errors.

4.2 Quantum Algorithms

The preceding section outlines plans to turn quantum effects into actual gates and, eventually, into quantum computers. But how much faster might quantum computers be? After all, last year's laptop computer seemed fast until this year's computer arrived. To succinctly address this, we introduce some complexity theory.

To help answer whether one algorithm or computer is actually faster we count the number of basic operations a program executes, not (necessarily) the execution, or elapsed "wall clock" time. Differences in elapsed time may be due to differences in the compiler, a neat programming trick, memory caching, or the presence of other running programs. We want to concentrate on fundamental differences, if any, rather than judging a programming competition.

In measuring algorithm performance we must consider the size of the input. A computer running a program to factor a 10,000-digit number shouldn't be compared with a different computer that is only factoring a 10-digit number. So we will compare performance in terms of the size of the problem or input. We expect that larger problems take longer to solve than smaller instances of the same problem. Hence, we express performance as a function of the problem size, e.g., $f(n)$. We will see that performance functions fall into theoretically neat and practically useful "complexity classes."

4.2.1 Complexity Classes

What are some typical complexity classes? Consider the problem of finding a name in a telephone book. If one took a dumb, but straightforward method where we check every entry, one at a time, from the beginning, the expected average number of checks is $n/2$ for a telephone book with n names. (This is called "sequential search.") Interestingly, if one checks names completely at random, even allowing accidental rechecks of names, the expected average number of checks is still $n/2$. Since telephone books are sorted by name, we can do much better. We estimate where the name will be, and open the book to that spot. Judging from the closeness to the name, we extrapolate again where the name will be and skip there. (This is called "extrapolation search.") This search is much faster, and on average takes some constant multiple of the logarithm of the number of names, or $c \log n$.

Although it takes a little longer to estimate and open the book than just check the next name, as n gets big, those constant multipliers don't matter. For large values of n the logarithm is so much smaller than n itself, it is clear that extrapolation search is far faster than linear search. (When one is close to the name, that is, when n is small, one switches to linear searching, since the time to do a check and move to the next name is much smaller.)

This is a mathematically clear distinction. Since almost any reasonable number times a logarithm is eventually smaller than another constant times the number, we'll ignore constant multiples (in most cases) and just indicate what "order" they are. We say that linear search is O(n), read "big-Oh of n", and extrapolation search is O($\log n$), read "big-Oh of log n." Since logarithms in different bases only differ by a constant multiple, we can (usually) ignore the detail of the logarithm's base.

Other common problems take different amounts of time, even by these high-level comparisons. Consider the problem of finding duplicates in an unordered list of names. Comparing every name to every other name takes some multiple of n^2, or O(n^2) time. Even if we optimize the algorithm and only compare each name to those after it, the time is still a multiple of n^2. Compared with O($\log n$) or even O(n), finding duplicates will be much slower than finding a name, especially for very large values of n.

It turns out we can sort names in time proportional to $n \log n$. Checking for duplicates in a *sorted* list only takes O(n), so sorting and then checking takes O($cn \log n + dn$), for some constants c and d. Since the $n \log n$ term is significantly bigger than the n term for large n, we can ignore the lone n and say this method is O($n \log n$), much faster than the O(n^2) time above.

Although the difference between these methods is significant, both are still *polynomial*, meaning the run time is a polynomial of the size. That is, they are both O(n^k) for some constant k. We find a qualitative difference in performance between polynomial algorithms and algorithms with *exponential* run time, that is, algorithms that are O(k^n) for some constant k. Polynomial algorithms are generally practical to run; even for large problems, the run time doesn't increase too much, whereas exponential algorithms are generally intractable. Even seemingly minor increases in the problem size can make the computation completely impractical.

There is a mathematical reason to separate polynomial from exponential algorithms. A polynomial of a polynomial is still a polynomial. Thus even having polynomial algorithms use other polynomial algorithms still results in a polynomial algorithm. Moreover any exponential function always grows faster than any polynomial function.

4.2.2 A Formal Definition of big-Oh Complexity

For the curious reader, we formally define big-Oh notation. We say $f(n) = O(g(n))$ if there are two positive constants k and n_0 such that $|f(n)| \leq kg(n)$ for all $n > n_0$. The constants k and n_0 must not depend on n. Informally, there is a constant k such that for large values (beyond n_0), $kg(n)$ is greater than $f(n)$.

From this definition, we see that constant multipliers are absorbed into the k. Also lower order terms, such as dn, are eventually dominated by higher order terms, like $cn \log n$ or n^2. Because a "faster" algorithm may have such large constants or lower order terms, it may perform worse than a "slower" algorithm for realistic problems. If we are clearly only solving small problems, it may, in fact, be better to use the "slower" algorithm, especially if the slower algorithm is simpler. However, experience shows that big-Oh complexity is usually an excellent measure to compare algorithms.

4.2.3 Shor's Factoring Algorithm

We can now succinctly compare the speed of computations. The security of a widely used encryption scheme, RSA, depends on the presumption that finding the factors of large numbers is intractable. After *decades* of work, the best classical factoring algorithm is the Number Field Sieve [LL93]. With it, factoring an n-digit number takes about $O\left(e^{\sqrt[3]{n}}\right)$ steps[5], which is exponential in n. What does this mean? Suppose you use RSA to encrypt messages, and your opponent buys fast computers to break your code. Multiplication by the Schönhage–Strassen algorithm [Schö82] takes $O(n \log n \log \log n)$ steps. Using a key eight times longer means multiplications, and hence encrypting and decrypting time, takes at most 24 times longer to run, for $n > 16$. However, the time for your opponent to factor the numbers, and hence break the code, increases to $e^{\sqrt[3]{8n}} = e^{2\sqrt[3]{n}} = \left(e^{\sqrt[3]{n}}\right)^2$. In other words, the time to factor is *squared*. It doesn't matter whether the time is in seconds or days: factoring was exponential. Without too much computational overhead you can increase the size of your key beyond the capability of any conceivable computer your opponent could obtain. At least, that was the case until 1994.

In 1994, Peter Shor invented a quantum algorithm for factoring numbers that takes $O\left(n^2 \log n \log \log n\right)$ steps. [Shor95] This is polynomial, and, in fact, isn't too much longer than the naïve time to multiply. So if you can encrypt, a determined opponent can break the code, as long as a quantum computer is available. With this breakthrough, the cryptography community in particular became very interested in quantum computing.

---

[5] More precisely, it takes $e^{\sqrt[3]{n(\log n)^2 64/9}}$ steps.

Shor's algorithm, like most factoring algorithms, uses "a standard reduction of the factoring problem to the problem of finding the period of a function." [RP00] What is the period of, say, the function $3^n \bmod 14$? Values of the function for increasing exponents are $3^1 = 3$, $3^2 = 9$, $3^3 = 27$ or $13 \bmod 14$, $3^4 = 11 \bmod 14$, $3^5 = 5 \bmod 14$, and $3^6 = 1 \bmod 14$. Since the function has the value 1 when $n=6$, the period of $3^n \bmod 14$ is 6.

The algorithm has five main steps to factor a composite number N with Shor's algorithm.
1. If N is even or there are integers a and $b > 1$ such that $N=a^b$, 2 or a are factors.
2. Pick a positive integer, m, which is relatively prime to N.
3. Using a quantum computer, find the period of $m^P \bmod N$, that is, the smallest positive integer P such that $m^P = 1 \bmod N$.
4. For number theoretic reasons, if P is odd or if $m^{P/2} + 1 = 0 \bmod N$, start over again with a new m at step 2.
5. Compute the greatest common divisor of $m^{P/2} - 1$ and N. This number is a divisor of N.

For a concrete example, let $N=323$, which is 19×17. N is neither even nor the power of an integer. Suppose we choose 4 for m in step 2. Since 4 is relatively prime to 323, we continue to step 3. We find that the period, P, is 36, since $4^{36} = 1 \bmod 323$. We do not need to repeat at step 4 since 36 is not odd and $4^{36/2} + 1 = 306 \bmod 323$. In step 5 we compute the greatest common divisor of $4^{36/2} - 1$ and 323, which is 19. Thus we have found a factor of 323.

The heart of Shor's algorithm is quantum period finding, which can also be applied to a quantum Fourier transform and finding discrete logarithms. These are exponentially faster than their classical counterparts.

4.2.4 Deutsch's Function Characterization Problem

To more clearly illustrate quantum computing's potential speedup, let's examine a contrived, but simple problem first presented and solved by Deutsch [Deu85]. Suppose we wish to find out whether an unknown boolean unary function is constant, either 0 or 1, or not. Classically, we must apply the function twice, once with a 0 input and once with a 1. If the outputs are both 0 or both 1, it is constant; otherwise, it is not. A single classical application of the function, say applying a 1, can't give us enough information. However, a single quantum application of the function can.

Using a superposition of 0 and 1 as the input, one quantum computation of the function yields the answer. The solution of Cleve, Ekert, Macchiavello and Mosca [CEMM98] to Deutsch's problem uses a common quantum computing operation, called a Hadamard[6], which converts $|0\rangle$ into the superposition $1/\sqrt{2}\left(|0\rangle+|1\rangle\right)$ and $|1\rangle$ into the superposition $1/\sqrt{2}\left(|0\rangle-|1\rangle\right)$. The algorithm is shown schematically in Fig. 9. The Hadamard is

---

[6] Also called Walsh transform, Walsh–Hadamard transform, or discrete Fourier transformation over Z_2^n.

represented as a box with an "H" in it. The function to be characterized is a box labeled "U$_f$".

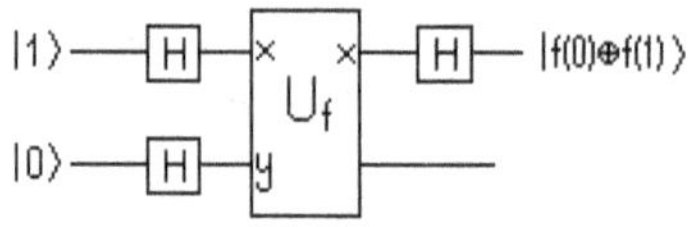

Figure 9. Solution to Deutsch's Function Characterization Problem

To begin, we apply a Hadamard to a $|0\rangle$ and another Hadamard to a $|1\rangle$.

$$H|0\rangle H|1\rangle = 1/2\,(|0\rangle + |1\rangle)(|0\rangle - |1\rangle)$$

$$= 1/2\,(|0\rangle(|0\rangle - |1\rangle) + |1\rangle(|0\rangle - |1\rangle))$$

To be reversible, the function, U_f, takes a pair of qubits, $|x\rangle|y\rangle$, and produces the pair $|x\rangle|y \oplus f(x)\rangle$. The second qubit is the original second qubit, y, exclusive-or'd with the function applied to the first qubit, f(x). We apply the function one time to the result of the Hadamards, and then apply another Hadamard to the first qubit, *not the "result" qubit*. Below, the "I" represents the identity, that is, we do nothing to the second qubit.

$$(H \otimes I)U_f\,1/2\,(|0\rangle(|0\rangle - |1\rangle) + |1\rangle(|0\rangle - |1\rangle))$$

$$= (H \otimes I)1/2\,(U_f|0\rangle(|0\rangle - |1\rangle) + U_f|1\rangle(|0\rangle - |1\rangle))$$

$$= (H \otimes I)1/2\,(|0\rangle(|0 \oplus f(0)\rangle - |1 \oplus f(0)\rangle) + |1\rangle(|0 \oplus f(1)\rangle - |1 \oplus f(1)\rangle))$$

$$= 1/2\,(H|0\rangle|0 \oplus f(0)\rangle - H|0\rangle|1 \oplus f(0)\rangle + H|1\rangle|0 \oplus f(1)\rangle - H|1\rangle|1 \oplus f(1)\rangle)$$

$$= 1/2\sqrt{2}\begin{pmatrix}(|0\rangle + |1\rangle)|0 \oplus f(0)\rangle - (|0\rangle + |1\rangle)|1 \oplus f(0)\rangle + \\ (|0\rangle - |1\rangle)|0 \oplus f(1)\rangle - (|0\rangle - |1\rangle)|1 \oplus f(1)\rangle\end{pmatrix}$$

Case analysis and algebraic manipulations reduce this equation to the following. (Details are in Appendix A.)

$$= 1/\sqrt{2}\,|f(0) \oplus f(1)\rangle(|0\rangle - |1\rangle)$$

$$= |f(0) \oplus f(1)\rangle \otimes 1/\sqrt{2}\,(|0\rangle - |1\rangle)$$

We now measure the first qubit. If it is 0, the function is a constant. If we measure a 1, the function is not a constant. Thus we can compute a property of a function's range using only one function application.

Although contrived, this example shows that using quantum entanglement and superposition, we can compute some properties faster than is possible with classical means. This is part of the lure of quantum computing research.

4.2.5 Grover's Search Algorithm

The requirement of searching for information is simple: find a certain value in a set of values with no ordering. For example, does the name "John Smith" occur in a set of 1,000 names? Since there is no order to the names, the best classical solution is to examine each name, one at a time. For a set of N names, the expected run time is $O(N/2)$: on average we must examine half the names.

There are classical methods to speed up the search, such as sorting the names and doing a binary search, or using parallel processing or associative memory. Sorting requires us to assign an order to the data, which may be hard if we are searching data such as pictures or audio recordings. To accommodate every possible search, say by last name or by first name, we would need to create separate sorted indices into the data, requiring $O(N \log N)$ preliminary computation and $O(N)$ extra storage. Parallel processing and associative memory takes $O(N)$ resources. Thus these classical methods speed up query time by taking time earlier or using more resources.

In 1996 Lov K. Grover presented a quantum algorithm [G96a, G96b] to solve the general search problem in $O(\sqrt{N} \log N)$ time. The algorithm proceeds by repeatedly enhancing the amplitude of the position in which the name occurs.

Database searching, especially for previously unindexed information, is becoming more important in business operations, such as data mining. However, Grover's algorithm might have an impact that reaches even farther. Although we present the algorithm in terms of looking for names, the search can be adapted to any recognizable pattern. Solutions to problems currently thought to take more than polynomial time, that is, $O(k^n)$, may be solvable in polynomial time. A typical problem in this group is the Traveling Salesman Problem, which is, finding the shortest path between all points in a set. This problem occurs in situations such as finding the best routing of trucks between pick up and drop off points, airplanes between cities, and the fastest path of a drill head making holes in a printed circuit board. The search algorithm would initialize all possible solutions, and then repeatedly enhance the amplitude of the best solution. No published quantum algorithm exists to solve the Traveling Salesman, or any NP problem, in polynomial time. However, improvements like Grover's hint that it may be possible.

4.2.6 Quantum Simulation

Since a quantum system is the tensor product of its component systems, the amount of information needed to completely describe an arbitrary quantum system increases exponentially with the size. This means that classical simulation of quantum systems with even a dozen qubits challenges the fastest supercomputers. Research into protein folding to discover new drugs, evaluating different physical models of the universe, understanding new superconductors, or designing quantum computers may take far more classical computer power than could reasonably be expected to exist on Earth in the next decade. However, since quantum computers can represent an exponential amount of information, they may make such investigations tractable.

4.3 Quantum Error Correction

One of the most serious problems for quantum information processing is that of decoherence, the tendency for quantum superpositions to collapse into a single, definite, classical state. As we have seen, the power of quantum computing derives in large part from the ability to take advantage of the unique properties of quantum mechanics—superposition and entanglement. The qubits that comprise a quantum computer must inevitably interact with other components of the system, in order to perform a computation. This interaction inevitably will lead to errors. To prevent the state of qubits from degrading to the point that quantum computations fail requires that errors be either prevented or corrected.

In classical systems, errors are prevented to some degree by making the ratio of system size to error deviation very large. Error correction methods are well known in conventional computing systems, and have been used for decades. Classical error correction uses various types of redundancy to isolate and then correct errors. Multiple copies of a bit or signal can be compared, with the assumption that errors are sufficiently improbable to never result in faulty bits or signals being more likely than valid ones, e.g., if three bits are used to encode a one-bit value, and two of three bits match, then the third is assumed to be faulty.

In quantum systems it is not possible to measure qubit values without destroying the superposition that quantum computing needs, so at first there was doubt that quantum error correction would ever be feasible. This is natural, especially considering the no-cloning theorem (Sect. 2.10). Not only could qubits not be exactly measured, they cannot even be arbitrarily copied by any conceivable scheme for detecting and correcting errors. It is perhaps surprising then that quantum error correction is not only possible, but also remarkably effective.

The challenge in quantum error correction is to isolate and correct errors without disturbing the quantum state of the system. It is in fact possible to use some of the same ideas employed for classical error correction in a quantum system; the trick is to match the redundancy to the type of errors likely to occur in the system. Once we know what kinds of errors are most likely, it is possible to design effective quantum error correction mechanisms.

4.3.1 Single Bit-flip Errors

To see how this is possible, consider a simple class of errors: single bit errors that affect qubits independently. (In reality, of course, more complex problems occur, but this example illustrates the basic technique.) Consider a single qubit, a two-state system with bases $|0\rangle$ and $|1\rangle$. We will use a simple "repetition code", that is, we represent a logical zero with three zero qubits, $|0_L\rangle = |000\rangle$, and a logical one with three ones, $|1_L\rangle = |111\rangle$. An arbitrary qubit in this system, written as a superposition $a|0_L\rangle + b|1_L\rangle$, becomes $a|000\rangle + b|111\rangle$ with repetition coding. Since we assume the system is stable in all ways except perhaps for single bit flips, there may be either no error or else one of three qubits flipped, as shown in Table 2.

The strategy for detecting errors is to add three "temporary" qubits $|t_0 t_1 t_2\rangle$, set to $|000\rangle$, which will hold "parity" results. We then XOR various bits together, putting the results in the added three qubits: t_0 is bit 0 XOR bit 1, t_1 is bit 0 XOR bit 2, and t_2 is bit 1 XOR bit 2. This leaves a unique pattern, called a syndrome, for each error. The third column of Table 2 shows the respective syndromes for each error.

Error	Error state	Syndrome	Correction
No error	$a|000\rangle + b|111\rangle$	$|000\rangle$	none
qubit 1 flipped	$a|100\rangle + b|011\rangle$	$|110\rangle$	$X \otimes I \otimes I$
qubit 2 flipped	$a|010\rangle + b|101\rangle$	$|101\rangle$	$I \otimes X \otimes I$
qubit 3 flipped	$a|001\rangle + b|110\rangle$	$|011\rangle$	$I \otimes I \otimes X$

Table 2. Bit-flip Errors, Syndromes, and Correctives.

Measuring the added three qubits yields a syndrome, *while maintaining the superpositions and entanglements we need*. Depending on which syndrome we find, we apply one of the three corrective operations given in the last column to the original three repetition encoding qubits. The operation X flips a bit, that is, it changes a 0 to a 1 and a 1 to a 0. The identity operation is I. We illustrate this process in the following example.

4.3.2 An Error Correction Example

In fact, the repetition code can correct a superposition of errors. This is more realistic than depending on an error affecting only one qubit. It also illuminates some quantum behaviors. Like any other quantum state, the error may be a superposition, such as the following:

$$(\sqrt{.8} X \otimes I \otimes I + \sqrt{.2} I \otimes X \otimes I)(a|000\rangle + b|111\rangle).$$

Informally the first factor may be read as, if we measured the state, we have an 80% chance of finding the first qubit flipped and a 20% chance of finding the second qubit flipped. Multiplying the error state is

$$|\Psi\rangle = (\sqrt{.8}X \otimes I \otimes I + \sqrt{.2}I \otimes X \otimes I)(a|000\rangle + b|111\rangle)$$

$$= \sqrt{.8}(a|100\rangle + b|011\rangle) + \sqrt{.2}(a|010\rangle + b|101\rangle).$$

The error state is then augmented with $|000\rangle$ and the syndrome extraction, S, applied:

$$S(|\Psi\rangle \otimes |000\rangle)$$

$$= S(\sqrt{.8}(a|100000\rangle + b|011000\rangle) + \sqrt{.2}(a|010000\rangle + b|101000\rangle))$$

$$= \sqrt{.8}(a|100110\rangle + b|011110\rangle) + \sqrt{.2}(a|010101\rangle + b|101101\rangle)$$

$$= \sqrt{.8}(a|100\rangle + b|011\rangle) \otimes |110\rangle + \sqrt{.2}(a|010\rangle + b|101\rangle) \otimes |101\rangle$$

Now we measure the last three qubits. This measurement collapses them to $|110\rangle$ with 80% probability or $|101\rangle$ with 20% probability. Since they are entangled with the repetition coding bits, the coding bits partially collapse, too. The final state is $(a|100\rangle + b|011\rangle) \otimes |110\rangle$ with 80% probability or $(a|010\rangle + b|101\rangle) \otimes |101\rangle$ with 20% probability. If we measured 1, 1, 0, the first collapse took place, and we apply $X \otimes I \otimes I$ to $a|100\rangle + b|011\rangle$ producing $a|000\rangle + b|111\rangle$, the original coding. On the other hand, if we measured 1, 0, 1, we apply $I \otimes X \otimes I$ to $a|010\rangle + b|101\rangle$. In either case, the system is restored to the original condition, $a|000\rangle + b|111\rangle$, without ever measuring (or disturbing) the repetition bits themselves.

This error correction model works only if no more than one of the three qubits experiences an error. With an error probability of p, the chance of either no error or one error is $(1-p)^3 + 3p(1-p)^2 = 1 - 3p^2 + 2p^3$. This method improves system reliability if the chance of an uncorrectable error, which is $3p^2 - 2p^3$, is less than the chance of a single error, p, in other words, if $p < 0.5$.

4.3.3 From Error Correction to Quantum Fault Tolerance

The replication code given above is simple, but has disadvantages. First, it only corrects "bit flips", that is, errors in the state of a qubit. It cannot correct "phase errors", such as the change of sign in $1/\sqrt{2}(|0\rangle + |1\rangle)$ to $1/\sqrt{2}(|0\rangle - |1\rangle)$. Second, a replication code wastes resources. The code uses three actual qubits to encode one logical qubit. Further improvements in reliability take significantly more resources. More efficient codes can correct arbitrary bit or phase errors while using a sublinear number of additional qubits.

One such coding scheme is *group codes*. Since the odds of a single qubit being corrupted must be low (or else error correction wouldn't work at all), we can economize by protecting a group of qubits at the same time, rather than protecting the qubits one at a time. In 1996 Ekert and Macchiavello pointed out [EM96] that such codes were possible and showed a lower bound. To protect *l* logical qubits from up to *t* errors, they must be encoded in the entangled state of at least *n* physical qubits, such that the following holds.

$$2^l \sum_{i=0}^{t} 3^i \binom{n}{i} \leq 2^n$$

An especially promising approach is the use of "concatenated" error correcting codes [KL96, KL97]. In this scheme, a single logical qubit is encoded as several qubits, but in addition the code qubits themselves are also encoded, forming a hierarchy of encodings. The significance is that if the probability of error for an individual qubit can be reduced below a certain threshold, then quantum computations can be carried out to an arbitrary degree of accuracy.

A new approach complements error correction. Fault tolerant quantum computing avoids the need to actively decode and correct errors by computing directly on encoded quantum states. Instead of computing with gates and qubits, fault tolerant designs use procedures that execute *encoded gates* on encoded states that represent logical qubits. Although many problems remain to be solved in the physical implementation of fault tolerant quantum computing, this approach brings quantum computing a little closer to reality.

5 Quantum Communication and Cryptography

Quantum computing promises a revolutionary advance in computational power, but applications of quantum mechanics to communication and cryptography may have equally spectacular results, and practical implementations may be available much sooner. In addition, quantum communication is likely to be just as essential to quantum computing as networking is to today's computer systems. Most observers expect quantum cryptography to be the first practical application for quantum communications and computing.

5.1 Why Quantum Cryptography Matters

Cryptography has a long history of competition between code makers and code breakers. New encryption methods appear routinely, and many are quickly cracked through lines of attack that their creators never considered. During the first and second World Wars, both sides were breaking codes that the other side considered secure. More significantly, a code that is secure at one time may fall to advances in technology. The most famous example of this may be the World War II German Enigma code. Some key mathematical insights made it possible to break Enigma messages encrypted with poorly selected keys, but only with an immense amount of computation. By the middle of the war, Enigma messages were being broken using electromechanical computers developed first by

Polish intelligence and later by faster British devices built under the direction of Alan Turing. Although the Germans improved their encryption machines, Joseph Desch, at NCR Corporation, developed code breaking devices 20 times faster than Turing's, enabling the US Navy's Op-20-G to continue cracking many Enigma messages. Today, an average personal computer can break Enigma encryption in seconds.

A quantum computer would have the same impact on many existing encryption algorithms. Much of modern cryptography is based on exploiting extremely hard mathematical problems, for which there are no known efficient solutions. Many modern cipher methods are based on the difficulty of factoring (see Section 4.2.3) or computing discrete logarithms for large numbers (e.g., over 100 digits). The best algorithms for solving these problems are exponential in the length of input, so a brute force attack would require literally billions of years, even on computers thousands of times faster than today's machines.

Quantum computers factoring large numbers or solving discrete logarithms would make some of the most widely used encryption methods obsolete overnight. Although quantum computers are not expected to be available for at least the next decade, the very existence of a quantum factoring algorithm makes classical cryptography obsolete for some applications. It is generally accepted that a new encryption method should protect information for 20 to 30 years, given expected technological advances. Since it is conceivable that a quantum computer will be built within the next two to three decades, algorithms based on factoring or discrete logarithms are, in that sense, obsolete already. Quantum cryptography, however, offers a solution to the problem of securing codes against technological advances.

5.2 Unbreakable Codes

An encrypted message can always be cryptanalyzed by brute force methods—trying every key until the correct one is found. There is, however, one exception to this rule. A cipher developed in 1917 by Gilbert Vernam of AT&T, is truly unbreakable. A Vernam cipher, or "one-time pad", uses a key with a random sequence of letters in the encryption alphabet, equal in length to the message to be encrypted. A message, M, is encrypted by adding, modulo the alphabet length, each letter of the key K to the corresponding letter of M, i.e., $C_i = M_i \oplus K_i$, where C is the encrypted message, or *ciphertext*, and $\oplus$ is modular addition (see Table 3). To decrypt, the process is reversed.

Text	Random key	Ciphertext
C (3)	$\oplus$ U (21)	X (24)
A (1)	$\oplus$ D (4)	E (5)
T (20)	$\oplus$ I (9)	C (3)

Table 3. One time pad

The Vernam cipher is unbreakable because there is no way to determine a unique match between encrypted message C and key K. Since the key is random and the same length as the message, an encrypted message can decrypt to any text at all, depending on the key that is tried. For example, consider the ciphertext 'XEC'. Since keys are completely random, all keys are equally probable. So it is just as likely that the key is 'UDI', which decrypts to 'CAT', or 'TPW', which decrypts to 'DOG'. There is no way to prove which is the real key, and therefore no way to know the original message.

Although it is completely secure, the Vernam cipher has serious disadvantages. Since the key must be the same length as the message, a huge volume of key material must be exchanged by sender and recipient. This makes it impractical for high-volume applications such as day-to-day military communication. However, Vernam ciphers may be used to transmit master keys for other encryption schemes. Historically, Vernam ciphers have been used by spies sending short messages, using pads of random keys that could be destroyed after each transmission, hence the common name "one-time pad."

An equally serious problem is that if the key is ever reused, it becomes possible to decrypt two or more messages that were encrypted under the same key. A spectacular example of this problem is the post-war decryption of Soviet KGB and GRU messages by US and British intelligence under the code name VENONA. Soviet intelligence had established a practice of reusing one-time pads after a period of years. British intelligence analysts noticed a few matches in ciphers from a large volume of intercepted Soviet communications [Wri87]. Over a period of years, British and US cryptanalysts working at Arlington Hall in Virginia gradually decrypted hundreds of Soviet messages, many of them critical in revealing Soviet espionage against US atomic weapons research in the 1940s and early 1950s.

Still another problem with implementing a Vernam cipher is that the key must be truly random. Using text from a book, for example, would not be secure. Similarly, using the output of a conventional cipher system, such as DES, results in an encryption that is only as secure as the cipher system, not an unbreakable one-time pad system. Pseudo-random number generator programs may produce sequences with correlations or the entire generation algorithm may be discovered; both these attacks have been successfully used. Thus while the Vernam cipher is in theory unbreakable, in practice it becomes difficult and impractical for most applications. Conventional cryptosystems, on the other hand, can be broken but are much more efficient and easier to use.

5.3 Quantum Cryptography

Quantum cryptography offers some potentially enormous advantages over conventional cryptosystems, and may also be the only way to secure communications against the power of quantum computers. With quantum methods, it becomes possible to exchange keys with the guarantee that any eavesdropper to intercept the key is detectable with arbitrarily high probability. If the keys are used as one-time pads, complete security is assured. Although special purpose classical hardware can generate keys that are truly random, it is easy to use the collapse of quantum superpositions to generate truly random keys. This eliminates one of the major drawbacks to using one-time pads. The ability to detect the presence of an eavesdropper is in itself a huge advantage over conventional methods. With ordinary cryptography, there is always a risk that the key has been intercepted. Quantum key distribution eliminates this risk using properties of quantum mechanics to reveal the presence of any eavesdropping.

5.3.1 Quantum Key Distribution

The first significant communications application proposed using quantum effects is quantum key distribution, which solves the problem of communicating a shared cryptographic key between two parties with complete security. Classical solutions to the key distribution problem all carry a small, but real, risk that the encrypted communications used for sharing a key could be decrypted by an adversary. Quantum key distribution (QKD) can, in theory, make it impossible for the adversary to intercept the key communication without revealing his presence. The security of QKD relies on the physical effects that occur when photons are measured.

As discussed in Section 3.3, a photon polarized in a given direction will not pass through a filter whose polarization is perpendicular to the photon's polarization. At any other angle than perpendicular, the photon may or may not pass through the filter, with a probability that depends on the difference between the direction of polarization of the photon and the filter. At 45°, the probability of passing through the filter is 50%.

The filter is effectively a measuring device. According to the measurement postulate of quantum mechanics, measurements in a 2-dimensional system are made according to an orthonormal basis[7]. Measuring the state transforms it into one or the other of the basis vectors. In effect, the photon is forced to "choose" one of the basis vectors with a probability that depends on how far its angle of polarization is from the two basis vectors. For example, a diagonally polarized photon measured according to a vertical/horizontal basis will be in a state of either vertical or horizontal polarization after measurement. Furthermore, any polarization angle can be represented as a linear combination, $a|\uparrow\rangle + b|\rightarrow\rangle$ of orthogonal (i.e., perpendicular) basis vectors. For QKD, two bases are used: rectilinear, with basis vectors ↑ and →, and diagonal, with basis vectors ↗ and ↖. Measuring photons in these polarizations according to the basis vectors produces the results shown in Table 4.

Polarization	↑	→	↗	↖	↑	→	↗	↖
Basis	+	+	+	+	X	X	X	X
Result	↑	→	↑ or → (50/50)	↑ or → (50/50)	↗ or ↖ (50/50)	↗ or ↖ (50/50)	↗	↖

Table 4. Photon measurement with different bases.

---

[7] Recall from linear algebra that a *basis* for a vector space is a set of vectors that can be used in linear combination to produce any vector in the space. A set of k vectors is necessary and sufficient to define a basis for a k-dimensional space. A commonly used basis for a 2-dimensional vector space is (1,0) and (0,1).

These results are the basis of a key distribution protocol devised by Bennett and Brassard [BB84]. Many other QKD protocols have been devised since, using similar ideas. Suppose two parties, Alice and Bob, wish to establish a shared cryptographic key. An eavesdropper, Eve, is known to be attempting to observe their communication; see Figure 10. How can the key be shared without Eve intercepting it? Traditional solutions require that the key be encrypted under a previously shared key, which carries the risk that the communication may be decrypted by cryptanalytic means, or that the previously shared key may have been compromised. Either way, Eve may read the message and learn Alice and Bob's new key. QKD provides a method to establish the shared key that guarantees that either the key will be perfectly secure, or that Alice and Bob will learn that Eve is listening and therefore not use the key.

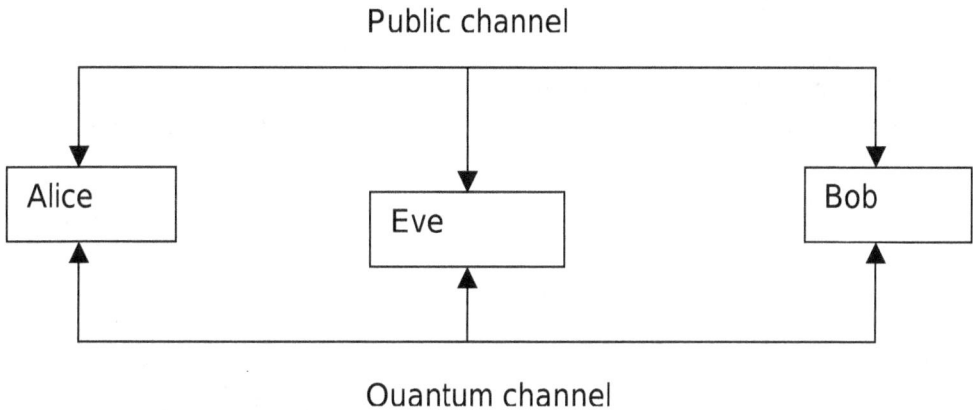

Figure 10. Quantum key distribution.

The BB84 QKD protocol takes advantage of the properties shown in Table 4. The protocol proceeds as follows:

Alice and Bob agree in advance on a representation for 0 and 1 bits in each basis. For example, they may choose → and ↗ to represent 0 and ↑ and ↖ to represent 1. Alice sends to Bob a stream of polarized photons, choosing randomly between ↑, →, ↗, and ↖ polarizations. When receiving a photon, Bob chooses randomly between + and x bases. When the transmission is complete, Bob sends Alice the sequence of bases he used to measure the photons. This communication can be completely public. Alice tells Bob which of the bases were the same ones she used. This communication can also be public. Alice and Bob discard the measurements for which Bob used a different basis than Alice. On average, Bob will guess the correct basis 50% of the time, and will therefore get the same polarization as Alice sent. The key is then the interpretation of the sequence of remaining photons as 0s and 1s. Consider the example in Table 5.

Sent by Alice	→	→	↑	↑	↗	↖	↗	→	→	↑	↖	↖	↖	↑	→	↗
Basis used by Bob	X	+	X	X	+	X	+	+	+	X	X	+	X	+	+	X
Bob's result	↖	→	↗	↖	→	↖	↑	→	→	↖	↖	→	↖	↑	→	↗
Key		0				1		0	0		1			1	0	0

Table 5. Deriving a new key.

Eve can listen to the messages between Alice and Bob about the sequences of bases they use and learn the bases that Bob guessed correctly. But this tells her nothing about the key, because Alice's polarizations were chosen randomly. If Bob guessed + as the correct polarization, Eve does not know whether Alice sent a → (0) or a ↑ (1) polarized photon, and therefore knows nothing about the key bit the photon represents.

What happens if Eve intercepts the stream sent by Alice and measures the photons? On average, Eve will guess the correct basis 50% of the time, and the wrong basis 50% of the time, just as Bob does. But when Eve measures a photon, its state is altered to conform to the basis Eve used, so Bob will get the wrong result in approximately half of the cases where he and Alice have chosen the same basis. Since they choose the same basis half the time, Eve's measurement adds an error rate of 25%. Consider the elaborated example in Table 6.

Sent by Alice	→	→	↑	↑	↗	↖	↗	→	→	↑	↖	↖	↖	↑	→	↗
Basis used by Eve	+	+	X	+	+	+	X	+	X	X	+	+	X	+	X	+
Eve's result	→	→	↗	↑	→	↑	↗	→	↖	↖	↑	→	↖	↑	↗	→
Basis used by Bob	X	+	X	X	+	X	+	+	+	X	X	+	X	+	+	X
Bob's result	↖	→	↗	↖	→	↗	↑	→	→	↖	↗	→	↖	↑	→	↗
Key		0			err		1	0	0		err			1	0	0

Table 6. Quantum key distribution with eavesdropping.

We describe details of real systems with some error rate and determining the error rate in Sect. 5.4.

5.3.2 Generating Random Keys

Properly implemented, the BB84 protocol guarantees that Alice and Bob share a key that can be used either as a one-time pad, or as a key for a conventional cryptographic algorithm. In either case, real security is only available if the key is truly random. Any source of non-randomness is a potential weakness that might be exploited by a cryptanalyst. This is one reason that ordinary pseudo-random number generator programs, such as are used for simulations, are hard to use for cryptography. Some conventional cryptosystems rely on special purpose hardware to generate random bits, and elaborate tests [NIST00] are used to ensure randomness.

One of the interesting aspects of quantum cryptography is that it provides a way to ensure a truly random key as well as allowing for detection of eavesdropping. Recall from Section 3.2 that for a superposition $a|\Psi\rangle + b|\Phi\rangle$, the probability of a measurement result of

Ψ is a^2, and of Φ is b^2. Therefore when a series of superpositions of $a|0\rangle + b|1\rangle$ is measured, 01 and 10 are measured with equal probability. Measuring a series of particles in this state therefore establishes a truly random binary sequence.

5.4 Prospects and Practical Problems

Although in theory the BB84 protocol can produce a guaranteed secure key, a number of practical problems remain to be solved before quantum cryptography can fulfill its promise. BB84 and other quantum protocols are idealized, but current technology is not yet close enough to the idealized description to implement quantum protocols as practical products. As of 2001, QKD has not been demonstrated over a distance of more than 50 kilometers, but progress has been steady. Commercial products using quantum protocols may be available by 2005, if problems in generating and detecting single photons can be overcome.

Single photon production is one of the greatest challenges for quantum communication. To prevent eavesdropping, transmission of one photon per time slot is needed. If multiple photons are produced in a time slot, it is possible for an adversary to count the number of photons without disturbing their quantum state. Then, if multiple photons are present, one can be measured while the others are allowed to pass, revealing key information without betraying the presence of the adversary. Current methods of generating single photons typically have an efficiency of less than 15%, leaving plenty of opportunity for Eve.

One method of dealing with noise problems is to use *privacy amplification* techniques. Whenever noise is present, it must be assumed that Eve could obtain partial information on the key bits, since it is not possible for Alice and Bob to know for certain whether the error rate results from ordinary noise or from Eve's intrusion. Privacy amplification distills a long key, about which Eve is assumed to have partial information, down to a much shorter key that eliminates Eve's information to an arbitrarily low level.

For privacy amplification, the first part of the protocol works exactly as before: Alice sends Bob qubits over a quantum channel, then the two exchange information over a public channel about which measurement bases they used. As before, they delete the qubits for which they used different measurement bases. Now, however, they also must delete bit slots in which Bob should have received a qubit, but didn't, either because of Eve's intrusion, or from *dark counts* at Bob's detector. Bob transmits the location of dark counts to Alice over the public channel.

Next, Alice and Bob publicly compare small parts of their raw keys to estimate the error rate, then delete these publicly disclosed bits from their key, leaving the *tentative final key*. If the error rate exceeds a pre-determined error threshold, indicating possible interception by Eve, they start over from the beginning to attempt a new key.

If the error rate is below the threshold, they remove any remaining errors from the rest of the raw key, to produce the *reconciled key* by using parity checks of sub-blocks of the tentative final key. To do this, they partition the key into blocks of length l such that each block is unlikely to contain more than one error. They each compute parity on all blocks and publicly compare results, throwing away the last bit of each compared block. If parity does not agree for a block, they divide the block into two, then compare parity on the sub-blocks, continuing in this binary search fashion until the faulty bit is found and deleted. This step is repeated with different random partitions until it is no longer efficient to continue. After this process, they select randomly chosen subsets of the remaining key, computing parity, discarding faulty bits and the last bit of each partition as before. This process continues for some fixed number of times to ensure with high probability that the key contains no error.

Because physical imperfections are inevitable in any system, it must be assumed that Eve may be able to obtain at least partial information. Eavesdropping may occur, even with significantly improved hardware, through either multiple photon splitting or by intercepting and resending some bits, but not enough to reveal the presence of the eavesdropper. To overcome this problem, Bennett, Brassard, Crépeau, and Maurer [BBCM95] developed a *privacy amplification* procedure that distills a secure key by removing Eve's information with an arbitrarily high probability.

During the privacy amplification phase of the protocol, Eve's information is removed. The first step in privacy amplification is for Alice and Bob to use the error rate determined above to compute an upper bound, b, on the number of bits in the remaining key that could be known to Eve. Using the number of bits in the remaining, reconciled key, n, and an adjustable security parameter s, they select n-k-s subsets of the reconciled key. The subset selection is done publicly, but the contents of the subsets are kept secret. Alice and Bob then compute parity on the subsets they selected, using the resulting parities as the final secret key. On average, Eve now has less than $2^{-s}/\ln 2$ bits of information about the final key.

Even if a reliable method of single photon production is developed, errors in transmission are as inevitable with quantum as with classical communication. Because quantum protocols rely on measuring the error rate to detect the presence of an eavesdropper, it is critical that the transmission medium's contribution to the error rate be as small as possible. If transmission errors exceed 25%, secure communication is not possible, because a simple man-in-the-middle attack—measuring all bits and passing them on— will not be detected.

5.5 Dense Coding

As discussed in Section 2.6, a qubit can produce only one bit of classical information. Surprisingly it is possible to communicate two bits of information using only one qubit and an EPR pair in a quantum technique known as *dense coding*. Dense coding takes advantage of entanglement to double the information content of the physically transmitted qubit.

Initially, Alice and Bob must each have one of the entangled particles of an EPR pair:

$$\Psi_0 = \frac{1}{\sqrt{2}}(|00\rangle + |11\rangle)$$

To communicate two bits, Alice represents the possible bit combinations as 0 through 3. Using the qubit in her possession, she then executes one of the transformations shown in Table 7.

Bits	Transform	New State
00	$\Psi_0 = (I \otimes I)\Psi_0$	$\frac{1}{\sqrt{2}}(|00\rangle + |11\rangle)$
01	$\Psi_1 = (X \otimes I)\Psi_0$	$\frac{1}{\sqrt{2}}(|10\rangle + |01\rangle)$
10	$\Psi_2 = (Z \otimes I)\Psi_0$	$\frac{1}{\sqrt{2}}(|00\rangle - |11\rangle)$
11	$\Psi_3 = (Y \otimes I)\Psi_0$	$\frac{1}{\sqrt{2}}(|01\rangle - |10\rangle)$

Table 7. Dense coding, phase 1.

After the transformation, Alice sends her qubit to Bob. Now that Bob has both qubits, he can use a controlled-NOT (prior to this, Alice and Bob could apply transformations only to their individual particles.) The controlled-NOT makes it possible to factor out the second bit, while the first remains entangled. See Table 8.

State	C-NOT Result	First Bit	Second Bit
$\frac{1}{\sqrt{2}}(|00\rangle + |11\rangle)$	$\frac{1}{\sqrt{2}}(|00\rangle + |11\rangle)$	$\frac{1}{\sqrt{2}}(|0\rangle + |1\rangle)$	$|0\rangle$
$\frac{1}{\sqrt{2}}(|10\rangle + |01\rangle)$	$\frac{1}{\sqrt{2}}(|11\rangle + |01\rangle)$	$\frac{1}{\sqrt{2}}(|0\rangle + |1\rangle)$	$|1\rangle$
$\frac{1}{\sqrt{2}}(|00\rangle - |11\rangle)$	$\frac{1}{\sqrt{2}}(|00\rangle - |10\rangle)$	$\frac{1}{\sqrt{2}}(|0\rangle - |1\rangle)$	$|0\rangle$
$\frac{1}{\sqrt{2}}(|01\rangle - |10\rangle)$	$\frac{1}{\sqrt{2}}(|01\rangle - |11\rangle)$	$\frac{1}{\sqrt{2}}(|0\rangle - |1\rangle)$	$|1\rangle$

Table 8. Deriving second bit.

Notice that after the controlled-NOT, it is possible to read off the values of the initial bits by treating $1/\sqrt{2}(|0\rangle + |1\rangle)$ as 0 and $1/\sqrt{2}(|0\rangle - |1\rangle)$ as 1. All that remains is to reduce the first qubit to a classical value by executing a Hadamard transform. The results are shown in Table 9.

First Bit	H(First Bit)
$\frac{1}{\sqrt{2}}(|0\rangle+|1\rangle)$	$\frac{1}{\sqrt{2}}\left(\frac{1}{\sqrt{2}}(|0\rangle+|1\rangle)+\frac{1}{\sqrt{2}}(|0\rangle-|1\rangle)\right)=\frac{1}{2}(|0\rangle+|1\rangle+|0\rangle-|1\rangle)=|0\rangle$
$\frac{1}{\sqrt{2}}(|0\rangle+|1\rangle)$	$\frac{1}{\sqrt{2}}\left(\frac{1}{\sqrt{2}}(|0\rangle+|1\rangle)+\frac{1}{\sqrt{2}}(|0\rangle-|1\rangle)\right)=\frac{1}{2}(|0\rangle+|1\rangle+|0\rangle-|1\rangle)=|0\rangle$
$\frac{1}{\sqrt{2}}(|0\rangle-|1\rangle)$	$\frac{1}{\sqrt{2}}\left(\frac{1}{\sqrt{2}}(|0\rangle+|1\rangle)-\frac{1}{\sqrt{2}}(|0\rangle-|1\rangle)\right)=\frac{1}{2}(|0\rangle+|1\rangle-|0\rangle+|1\rangle)=|1\rangle$
$\frac{1}{\sqrt{2}}(|0\rangle-|1\rangle)$	$\frac{1}{\sqrt{2}}\left(\frac{1}{\sqrt{2}}(|0\rangle+|1\rangle)-\frac{1}{\sqrt{2}}(|0\rangle-|1\rangle)\right)=\frac{1}{2}(|0\rangle+|1\rangle-|0\rangle+|1\rangle)=|1\rangle$

Table 9. Deriving first bit.

The dense coding concept can also be implemented using three qubits in an entangled state known as a GHZ state [GZTY00, Cer01]. With this procedure, Alice can communicate three bits of classical information by sending two qubits. Using local operations on the two qubits, Alice is able to prepare the GHZ particles in any of the eight orthogonal GHZ states. Through entanglement, her operations affect the entire three-state system, just as her operation on one qubit of an entangled pair changes the state of the two qubits in the pair. Similar to two-qubit dense coding, Bob measures his qubit along with the qubits received from Alice to distinguish one of the eight possible states encoded by three bits.

5.6 Quantum Teleportation

As shown in Sect. 3.6, it is impossible to clone, or copy, an unknown quantum state. However, the quantum state can be moved to another location using classical communication. The original quantum state is reconstructed exactly at the receiver, but the original state is destroyed. The No-Cloning theorem thus holds because in the end there is only one state. Quantum teleportation can be considered the dual of the dense coding procedure: dense coding maps a quantum state to two classical bits, while teleportation maps two classical bits to a quantum state. Both processes would not be possible without entanglement.

Initially, Alice has a qubit in an unknown state, i.e.,

$$\Phi = a|0\rangle + b|1\rangle$$

As with dense coding, Alice and Bob each have one of an entangled pair of qubits:

$$\Psi_0 = \frac{1}{\sqrt{2}}(|00\rangle + |11\rangle)$$

The combined state is

$$\Phi \otimes \Psi_0 = \left(a|0\rangle \otimes \frac{1}{\sqrt{2}}(|00\rangle + |11\rangle) + b|1\rangle \otimes \frac{1}{\sqrt{2}}(|00\rangle + |11\rangle) \right)$$

$$= \frac{1}{\sqrt{2}}(a|0\rangle|00\rangle + a|0\rangle|11\rangle) + \frac{1}{\sqrt{2}}(b|1\rangle|00\rangle + b|1\rangle|11\rangle)$$

$$= \frac{1}{\sqrt{2}}(a|000\rangle + a|011\rangle + b|100\rangle + b|111\rangle)$$

At this point, Alice has the first two qubits and Bob has the third. Alice applies a controlled-NOT next:

$$(CNOT \otimes I)(\Phi \otimes \Psi) = \frac{1}{\sqrt{2}}(a|000\rangle + a|011\rangle + b|110\rangle + b|101\rangle)$$

Applying a Hadamard, $H \otimes I \otimes I$, then produces:

$$\frac{1}{2}(a(|000\rangle + |100\rangle + |011\rangle + |111\rangle) + b(|010\rangle - |110\rangle + |001\rangle - |101\rangle))$$

The objective is to retrieve the original state, so we rewrite the state to move the amplitudes a and b through the terms. This shows it is possible to measure the first two bits, leaving the third in a superposition of a and b.

$$\frac{1}{2}(|00\rangle a|0\rangle + |10\rangle a|0\rangle + |01\rangle a|1\rangle + |11\rangle a|1\rangle + |01\rangle b|0\rangle - |11\rangle b|0\rangle + |00\rangle b|1\rangle - |10\rangle b|1\rangle)$$

$$= \frac{1}{2}(|00\rangle(a|0\rangle + b|1\rangle) + |01\rangle(a|1\rangle + b|0\rangle) + |10\rangle(a|0\rangle - b|1\rangle) + |11\rangle(a|1\rangle - b|0\rangle))$$

What does this equation mean? When we measure the first two qubits, we get one of the four possibilities. Since the third qubit is entangled, it will be in the corresponding state. For instance, if we measure $|00\rangle$, the system has collapsed to the first state, and the third qubit is the state $a|0\rangle + b|1\rangle$. Table 10 lists the four possible results. In each case, we can recover the original quantum state by applying a transform determined by the first two qubits, leaving the third qubit in the state $\phi = a|0\rangle + b|1\rangle$.

Left qubits	3rd qubit state	*Transform*	3rd qubit new state
$|00\rangle$	$a|0\rangle + b|1\rangle$	I	$a|0\rangle + b|1\rangle$
$|01\rangle$	$a|1\rangle + b|0\rangle$	X	$a|0\rangle + b|1\rangle$

$\|10\rangle$	$a\|0\rangle - b\|1\rangle$	Z	$a\|0\rangle + b\|1\rangle$
$\|11\rangle$	$a\|1\rangle - b\|0\rangle$	Y	$a\|0\rangle + b\|1\rangle$

Table 10. Quantum teleportation.

Just as quantum computing has required the development of new computing models beyond recursive function theory and Turing machines, quantum communication, teleportation, and dense coding show the need for new models of information beyond classical Shannon information theory.

6 Physical Implementations

Until now, we have talked primarily about theoretical possibilities and in general physical and mathematical terms. To actually build a quantum computer a specific physical implementation will be required. Several experimental systems have demonstrated that they can manipulate a few qubits, and in some restrictive situations very simple quantum calculations with limited fidelity have been performed. But, a reliable, useful quantum computer is still far in the future. Moreover, we should recall that the first classical computers were large mechanical machines—not electronic computers based on transistors on silicon chips. Similarly, it is likely that the technology used to build the first useful quantum computer will be very different from the technology that eventually wins out.

There are many different physical implementations that may satisfy all the necessary elements required for building a scalable quantum processor, but at the moment numerous technical and engineering constraints remain to be overcome. In this section we will list the required properties that a successful physical implementation must have along with several different physical implementations that may possess these properties.

6.1 General Properties Required to Build a Quantum Computer

We will first describe briefly the general physical traits that any specific physical implementation must have. Currently, there is a big gap between demonstrations in the laboratory and generally useful devices. Moreover, most proposed laboratory implementations and the experiments carried out to date fail to completely satisfy all of the general characteristics. Which system ultimately satisfies all these constraints at a level required to build a true extensible quantum processor is not known. Here, we will comment on the general properties needed to build a useful quantum computing device.

6.1.1 Well-Characterized Qubits

The first requirement is that the qubits chosen must be well characterized. This requires that each individual qubit must have a well-defined set of quantum states that will make up the "qubit". So far, we have assumed that we had a two level, or binary, quantum system. In reality, most quantum systems have more than two levels. In principle we could build a quantum computer whose individual elements or qubits consist of systems with four, five, ten, or any number of levels. The different levels we use may be a subset of the total number of levels in individual elements. Whatever the level structure of my qubit, we require that the levels being used have well defined properties, such as energy. Moreover, a superposition of the levels of the qubit must minimize decoherence by hindering energy from moving into or out of the qubit.

This general constraint requires that each individual qubit have the same internal level structure, regardless of its local external environment. This also requires that the qubit is well isolated from its environment to hinder energy flow between the environment and the qubit. Isolating information from the environment is easier for classical bits than quantum bits. A classical bit or switch is either in the state "0" or the state "1"; on or off. Except in special cases, such as communication, CCDs, or magnetic disks, we engineer classical systems to be in one of these two possibilities, never in between. In those special cases interactions are kept to a few tens of devices.

A quantum system or qubit is inherently a much more delicate system. Although we may prepare the system in some excited state $|1\rangle$, most quantum systems will decay to $|0\rangle$ or an arbitrary superposition because of interactions with the environment. Interaction with the environment must be controllable to build a large quantum processor. Moreover, in some proposed physical implementations, individual qubits may have slightly different internal level structure resulting either from the manufacturing process or the interaction of the qubit with its environment. This slight difference in level structure must be compensated and should not change during the computation.

The physical nature of the qubit may be any one of a number of properties, such as, electron spin, nuclear spin, photon polarization, the motional or trapping state of a neutral atom or ion, or the flux or charge in a superconducting quantum interference device (SQUID). For instance, in the particular case of ion traps, it is the ground state of the hyperfine states that result from actually coupling the electron and nuclear spin together. Further, it is only a specific pair of the magnetic sublevels of those hyperfine states. In a quantum dot one again uses the complex structure of the device to come up with two states to act as the qubit. In this case, it corresponds more to an excitation of an electron or an electron-hole pair.

6.1.2 Scalable Qubit Arrays and Gates

This requirement is a logical extension of the previous requirement. Since individual qubits must be able to interact to build quantum gates, they must be held in some type of replicated trap or matrix. In the very early days of computing mechanical switches or tubes in racks held information instead of microscopic transistors etched in silicon and mounted in integrated circuits. Thus the specific nature of the supporting infrastructure depends on the nature of the qubit. Regardless of the matrix or environment holding individual qubits, it is essential that we can add more qubits without modifying the properties of the previous qubits and having to reengineer the whole system. Because quantum systems are so sensitive to their environment, scalability is not trivial.

Scalability also requires that qubits are stable over periods that are long compared to both single qubit operations and two-qubit gates. In other words, the states of the qubits must not decohere on a time scale that is long compared to one and two qubit operations. This is increasingly important in larger quantum computers where larger portions of the computer must wait for some parts to finish.

6.1.3 Stability and speed

Physical implementations of a qubit are based on different underlying physical effects. In general these physical effects have very different decoherence times. For example, nuclear spin relaxation can be from one-tenth of a second to a year, whereas the decoherence time is more like 10^{-3} seconds in the case of electron spin. It is approximately 10^{-6} s for a quantum dot, and around 10^{-9} s for an electron in certain solid state implementations. Although one might conclude that nuclear spins are the best, decoherence time is not the only concern.

A qubit must interact with external agencies so two qubits can interact. The stronger the external interactions, the faster two-qubit gates could operate. Because of the weak interaction of the nuclear spin with its environment, gates will likely take from 10^{-3} to 10^{-6} seconds to operate, giving "clock speeds" of from 1 kHz to 1 MHz. The interaction for electron spin is stronger, so gates based on electron spin qubits may operate in 10^{-6} to 10^{-8} seconds or at 1 to 100 MHz. Thus electron spin qubits are likely to be faster, but less stable.

Since quantum error correction, presented in Sect. 4.3, requires gates like those used for computations, error correction only helps if we can expect more than about 10,000 operations before decoherence, that is, an error. Reaching this level of accuracy with a scalable method is the current milestone. Therefore it is really the ratio of gate operation time to decoherence time, or operations until decoherence, that is the near term goal.

Dividing decoherence times by operation times we may have from 10^2 to 10^{13} nuclear spin operations before a decoherence or between 10^3 and 10^5 electron spin operations before a decoherence. Although decoherence and operation times are very different, the number of operations may be similar. This is not surprising since, in general, the weaker the underlying interactions, the slower the decoherence and the slower the one- and two-qubit operations. We see that many schemes offer the possibility of more than 10,000 operations between decoherence events. The primary problem is engineering the systems to get high gate speeds and enough operations before decoherence.

It is not clear which physical implementation will provide the best qubits and gates. If we examine the state of the art in doing one-qubit operations while controlling decoherence, ions in an ion trap and single nuclear spins look the most promising.

However the weak interaction with the environment, and thus potentially other qubits, makes 2-qubit gates significantly slower than some solid-state implementations.

6.1.4 Good Fidelity

Fidelity is a measurement of the decoherence or decay of my qubits relative to one and two qubit gate times. Another aspect of fidelity is that when we perform an operation such as a CNOT, we do not expect to do it perfectly, but nearly perfectly. As an example, when we flip a classical bit from "0" to "1", it either succeeds or it fails—there is no in between. Even at a detailed view, it is relatively straightforward to strengthen classical devices to add more charge, increase voltage, etc., driving the value to a clean "0" or "1", charged or uncharged, on or off state. When we flip a quantum bit, we intend to exchange the amplitudes of the "0" and "1" states: $\alpha|0\rangle + \beta|1\rangle \rightarrow \beta|0\rangle + \alpha|1\rangle$. Since most physical effects we use are continuous values, the result of the operation is likely to have a small distortion: $\alpha|0\rangle + \beta|1\rangle \rightarrow \beta'|0\rangle + e^{i\varepsilon}\alpha'|1\rangle$. The error, ε, is nearly zero, and the primed quantities are almost equal to the unprimed quantities. But they are not perfectly equal. In this case, we require that the overlap between our expected result and the actual result be such that the net effect is to have a probability of error for a gate operation on the order of 10^{-4}.

6.1.5 Universal family of unitary transformations

In general, to build a true quantum computer it is only necessary to be able to perform an arbitrary one-qubit operations and *almost any* single two-qubit gate. If one can do arbitrary single qubit operations and almost any single two-qubit gate, one can combine these operations to perform single qubit operations, such as the Hadamard, and multi-qubit operations, such as a CNOT or C2NOT gate (see Sect. 4.1). [DBE95] From these, we know we can construct any boolean function.

6.1.6 Initialize values

Another important operation is the initialization of all the qubits into a well defined and characterized initial state. This is essential if one wishes to perform a specific algorithm since the initial state of the system must typically be known and be un-entangled. The initializing of the qubits corresponds to putting a quantum system into a completely coherent state that basically requires removing all thermal fluctuations and reduces the entropy (lack of order) of the system to zero. This is an extremely difficult task.

6.1.7 Readout

Another important requirement is the ability to reliably read resultant qubits. In many experimental situations, this is a technically challenging problem because one needs to either detect a quantum state of a system that has been engineered to only weakly interact with its environment. But this same system at "readout time" must interact sufficiently strongly that we can ascertain whether it is in the state $|0\rangle$ or $|1\rangle$, while simultaneously insuring that the result is not limited by our measurement or detection efficiency. Readout, along with single-qubit gates, implies we need to be able to uniquely address each qubit.

6.1.8 Types of Qubits

We must be able to store quantum information for relatively long times: the equivalent of main memory, or RAM, in classical computers. There are two general possibilities: material qubits, such as atoms or electrons, and "flying" qubits, or photons. Each has its own strengths and weaknesses. Material qubits can have decoherence times on the order of days, while photons move very fast and interact weakly with their environment. A quantum memory will likely be made of material systems consisting of neutral atoms or ions held in microtraps, solid state materials involving electron or nuclear spins, or artificial atoms like quantum dots. These material qubits are ideal for storing information if decoherence can be controlled. For example, single ions can be coherently stored for several days. However, manipulating individual photons or trying to build a two-qubit gate using photons appears quite difficult. A quantum processor is likely to use material qubits, too, to build the equivalent of registers and to interact well with the quantum memory.

6.1.9 Communication

When we wish to transmit or move quantum information, we typically want to use photons: they travel very fast and interact weakly with their environment. To build a successful quantum communication system will likely require the ability to move quantum information between material qubits and photons. This is another relatively difficult task, but several experiments have been successfully performed. However, different implementations of material qubits will likely need different solutions to moving entangled or superposed information from particles to photons and back.

6.2 Realizations

Just as there are many subatomic properties that may be exploited for quantum effects, realizations range from brand new technologies to decades old technologies harnessed and adapted for quantum computing. These technologies can be categorized into two basic classes. First, a top down approach where we take existing technology from the material science and solid state fields and adapt it to produce quantum systems. This top down approach involves creative ideas such as: implementing single ion impurities in silicon; designing very uniform quantum dots whose electronic properties are well characterized and controllable; using superconducting quantum interference devices, and several others.

A second contrasting approach is a bottom up approach. The idea here is to start with a good, natural qubit, such as an atom or ion, and trap the particle in a benign environment. This latter concept provides very good single qubits but leaves open the question of scalability—especially when one begins to examine the mechanical limits of current traps. The benefit of this approach is excellent, uniform, decoherence-free qubits with great readout and initialization capabilities. The hard problem will be scaling these systems and making the gate operations fast.

The top down approach suffers from decoherence in some cases or a dramatic failure of uniformity in the individual qubits: "identical qubits" are not truly uniform, decoherence-free individual qubits. The bottom up approach has the benefit of good quality qubits and starting with a basic understanding of decoherence processes. Below we will briefly discuss some of these possible technologies.

6.2.1 Charged Atoms in an Ion Trap

The one system that has had great success is ions in an ion trap. Dave Wineland's group at NIST, Boulder has
- entangled four ions,
- shown exceedingly long coherence times for a single qubit,
- demonstrated high efficiency readout,
- initialized four atoms into their ground state, and
- multiplexed atoms between two traps.

They have also shown violations of the Bell's inequalities [Bell64] and had many other successes. Their remarkable success and leadership of this effort blazes new frontiers in the experimental approaches to quantum computation, and their progress shows no signs of slowing.

Ions of beryllium are held single file. Laser pulses flip individual ions. To implement a CNOT gate, the motion of the ions "sloshing" back and forth in the trap is coupled to the electron levels. That is, if ions are in motion, the electron level is flipped. Otherwise the electron level is unchanged. This is the operation described abstractly in Sect. 4.1.

6.2.2 Neutral Atoms in Optical Lattices or Microtraps

Several groups are attempting to repeat the ion trap success using neutral atoms in optical lattices, where a trapping potential results from intersecting standing waves of light from four or more laser beams in free space, or micro-magnetic or micro-optical traps. These efforts are just getting seriously underway and appear to have many of the advantages of the ion scheme. One major difference is that the atoms interact less strongly with each other than with the ions. This could lead to better decoherence, but is also likely to lead to slower 2-qubit gate operations because of the weaker interactions. Much of the promise of the neutral atom approach is based on the remarkable advances made in the past two decades in laser cooling of atoms and the formation of neutral atom Bose-Einstein condensates where thousands of atoms are "condensed into" a single quantum state with temperatures of a few nanokelvins, or billionth of a degree above absolute zero. These advances have allowed scientist to manipulate large number of atoms in extremely controlled and exotic ways. The tools, techniques, and understandings developed over these past two decades may prove very useful in these current attempts to create quantum gates with these systems.

6.2.3 Solid State

Many different approaches fall under the realm of solid state. One general approach is to use quantum dots—so-called artificial atoms—as qubits. If it can be done in a controlled and decoherence free way, then one has the advantages of the atom and ion approach while having the controlled environment and assumed scalability that comes with solid state material processing. Another variant is embedding single atom impurities, such as ^{31}P, in silicon. The ^{31}P nuclear spin serves as a qubit while using basic semiconductor technology to build the required scalable infrastructure. Alternative approaches based on excitons in quantum dots or electronic spins in semiconductors are also being investigated. The primary difficulty of these approaches is building the artificial atoms or implanting the ^{31}P impurities precisely where required.

6.2.4 NMR

Nuclear magnetic resonance (NMR) has shown some remarkable achievements in quantum computing. However, it is widely believed that the current NMR approach will not scale to systems with more than 15 or 20 qubits. NMR uses ingenious series of radio frequency pulses to manipulate the nuclei of atoms in molecules. Although all isolated atoms of a certain element resonant at the same frequency, their interactions with other atoms in a molecule cause slight changes in resonance. The NMR approach is extremely useful in coming up with a series of pulses to manipulate relatively complex systems atoms in a molecule in situations where individual qubit rotations or gates might appear problematic. Thus this work provides useful knowledge into how to manipulate complex quantum systems. Low temperature solid state NMR is one possible way forward. Like the previous section, a single atom impurity—such as ^{31}P in Silicon—is a qubit, but NMR attempts to perform single site addressability, detection, and manipulation on nuclear spins.

6.2.5 Photon

Photons are clearly the best way to transmit information, since they move at the speed of light and do not strongly interact with their environment. This near-perfect characteristic for quantum communication makes photons problematic for quantum computation. In fact, early approaches to using photons for quantum computation suffered from a requirement of exponential number of optical elements and resources as one scaled the system. A second problem was that creating conditional logic for two-qubit gates appeared very difficult since two photons do not interact strongly even in highly nonlinear materials. In fact, most nonlinear phenomena involving light fields result only at high intensity.

Recently, new approaches for doing quantum computation with photons have appeared that depend on using measurement in a "dual-rail" approach to create entanglement. This approach removes many of the constraints of early approaches, but provides an alternative approach to creating quantum logic. Experimental efforts using this approach are just beginning. The approach will still have to solve the technically challenging problems caused by high-speed motion of their qubits, a benefit in communication and a possible benefit in computational speed, and by the lack of high efficient, single photon detectors that are essential to the success of this approach.

6.2.6 Optical Cavity Quantum Electrodynamics

Other atomic type approaches involve strongly coupling atoms or ions to photons using high finesse optical cavities. Similar type of approaches may be possible using tailored quantum dots, ring resonators, or photonic materials. One advantage of these types of approaches is the ability to move quantum information from photons to material qubits and back. This type of technology appears to be essential anyway since material qubits (e.g., atoms, ions, electrons, etc.) are best for storing quantum information while photons, i.e., flying qubits, are best for transmitting quantum information. It is possible that these approaches may provide very fast quantum processors as well. Numerous efforts are underway to investigate these schemes.

6.2.7 Superconducting Qubits

Superconducting quantum interference devices (SQUID) can provide two types of qubits: either flux based qubits, corresponding to bulk quantum circulation, or charge based qubits responsible for superconductivity. SQUID based science has been a field of investigation for several decades but have only recently shown an ability to observe Rabi-flopping—a key experiment that shows the ability to do single qubit operations. This approach to quantum computation has great potential but also will have to overcome numerous technical difficulties. One major issue is the need to operate a bulk system at liquid helium temperatures.

In summary, numerous physical approaches to quantum computing have been proposed and many are under serious research. Which of these approaches will ultimately be successful is not clear. In the near term, the ions and atomic systems will likely show the most progress, but the final winner will be the system that meets all of the technical requirements. This system may not even be among those listed above. What is important is that each of these approaches is providing us with an increased understanding of complex quantum systems and their coupling to the environment. This knowledge is essential to tackling the broad range of technical barriers that will have to be overcome to bring this exciting, perhaps revolutionary, field to fruition.

7 Conclusions

It will be at least a decade, and probably longer, before a practical quantum computer can be built. Yet the introduction of principles of quantum mechanics into computing theory has resulted in remarkable results already. Perhaps most significantly, it has been shown that there are functions that can be computed on a quantum computer that cannot be effectively computed with a conventional computer (i.e., a classical Turing machine.) This astonishing result has changed the understanding of computing theory that has been accepted for more than 50 years. Similarly, the application of quantum mechanics to information theory has shown that the accepted Shannon limit on the information carrying capacity of a bit can be exceeded. The field of quantum computing has produced a few algorithms that vastly exceed the performance of any conventional computer algorithm, but it is unclear whether these algorithms will remain rare novelties, or if quantum methods can be applied to a broad range of computing problems. The future of quantum communication is less uncertain, but a great deal of work is required before quantum networks can enter mainstream computing. Regardless of the future of practical quantum information systems, the union of quantum physics and computing theory has developed into a rich field that is changing our understanding of both computing and physics.

References

Many papers and articles on quantum computing are archived by Los Alamos National Laboratory with support by the United States National Science Foundation and Department of Energy. The URL for the quantum physics archive is http://arXiv.org/archive/quant-ph/. References to it are "quant-ph/YYMMNNN" where the date first submitted is given as YY (year), MM (month), NNN (number within month).

[Bell64] John Stewart Bell, "On the Einstein-Podolsky-Rosen Paradox", *Physics*, Vol. 1 (1964). pp. 195–200.

[BB84] Charles H. Bennett and Gilles Brassard, in Proc. IEEE International Conf. on Computers, Systems and Signal Processing, Bangalore, India, (IEEE, New York, 1984), p. 175.

[BBCM95] Charles H. Bennett, Gilles Brassard, Claude Crépeau, and Ueli M. Maurer. Generalized privacy amplification. *IEEE Transactions on Information Theory*, 41(6):1915–1923, November 1995.

[Cer01] Jose L. Cereceda, Quantum dense coding using three qubits, *C/Alto del Leon 8, 4A, 28038 Madrid, Spain*. May 21, 2001, quant-ph/0105096.

[CEMM98] Richard Cleve, Artur K. Ekert, Chiarra Macchiavello, and Michele Mosca, "Quantum Algorithms Revisited", *Proc. of Royal Society London*, A, 454:339–354, 1998.

[Deu85] David Deutsch, "Quantum Theory, the Church-Turing Principle and the Universal Quantum Computer", *Proc. of Royal Society London*, A, 400:97–117, 1985.

[DBE95] David Deutsch, Adriano Barenco, and Artur Ekert, "Universality in Quantum Computation", *Proc. of Royal Society London*, A, 499:669–677, 1995.

[EM96] Artur K. Ekert and Chiarra Macchiavello, "Quantum Error Correction for Communication", Physical Review Letters, 77:2585–2588, 1996, quant-ph/9602022.

[Feyn82] Richard Feynman, "Simulating Physics with Computers", *International Journal of Theoretical Physics* 21, 6&7, 467–488.

[GZTY00] V. N. Gorbachev, A. I. Zhiliba, A. I. Trubilko, and E. S. Yakovleva, *Teleportation of entangled states and dense coding using a multiparticle quantum channel*, quant-ph/0011124.

[G96a] Lov K. Grover, *A fast quantum mechanical algorithm for database search*, in 28th *Annual ACM Symposium on the Theory of Computation*, pages 212–219, ACM Press, New York, May 1996.

[G96b] Lov K. Grover, *A fast quantum mechanical algorithm for database search*, quant-ph/9605043.

[LL93] A. K. Lenstra and H. W. Lenstra, Jr., eds., *The Development of the Number Field Sieve*, Lecture Notes in Mathematics, Vol. 1554, Springer-Verlag, (1993).

[KL96] Emanuel Knill and Raymond LaFlamme, *Concatenated Quantum Codes*, quant-ph/9608012.

[KL97] Emanuel Knill and Raymond LaFlamme, "A Theory of Quantum Error-Correcting Codes", *Phys. Rev. A*, 55, 900 (1997).

[NIST00] National Institute of Standards and Technology, "A Statistical Test Suite for Random and Pseudorandom Number Generators for Cryptographic Applications", SP 800-22 October 2000.

[NC00] Michael A. Nielsen and Isaac L. Chuang, "Quantum Communication and Quantum Information", Cambridge University Press, 2000.

[RP00] Eleanor Rieffel and Wolfgang Polak, An Introduction to Quantum Computing for Non-Physicists, ACM Computing Surveys, 32(3):300–335, September 2000 quant-ph/9809016.

[Schö82] A. Schönhage, *Asympototically fast algorithms for the numerical multiplication and division of polynomials with complex coefficients*, in Computer Algebra EUROCAM '82, Lectures Notes in Computer Science, Vol. 144, pages 3–15, Springer-Verlag, 1982.

[Shor95] Peter W. Shor, *Polynomial time algorithms for prime factorization and discrete logarithms on a quantum computer*, SIAM J. on Computing, 26(5):1484–1509, 1997 quant-ph/9508027.

[You07] Thomas Young, "A Course of Lectures on Natural Philosophy and the Mechanical Arts", 1807. Johnson Reprint Corp., New York.

[Wri87] Peter Wright, "Spy Catcher: The Candid Autobiography of a Senior Intelligence Officer", Viking Penguin Inc, New York, 1987.

Appendix A.

The jump in the derivation of the answer to Deutsch's function characterization problem in Sect. 4.2.4 started after the application of the Hadamard, leaving us with the following equation.

$$1/2\sqrt{2}\begin{pmatrix}(|0\rangle+|1\rangle)|0\oplus f(0)\rangle - (|0\rangle+|1\rangle)|1\oplus f(0)\rangle + \\ (|0\rangle-|1\rangle)|0\oplus f(1)\rangle - (|0\rangle-|1\rangle)|1\oplus f(1)\rangle\end{pmatrix}$$

The easiest way to follow the result is to do case analysis. Here we have four cases: each possible result of $f(0)$ and $f(1)$. The exclusive-or operation is $0\oplus a = a$ and $1\oplus a = \bar{a}$.

Case I: $f(0)=0 \quad f(1)=0$

$$1/2\sqrt{2}\begin{pmatrix}(|0\rangle+|1\rangle)|0\oplus f(0)\rangle - (|0\rangle+|1\rangle)|1\oplus f(0)\rangle + \\ (|0\rangle-|1\rangle)|0\oplus f(1)\rangle - (|0\rangle-|1\rangle)|1\oplus f(1)\rangle\end{pmatrix}$$

$$= 1/2\sqrt{2}\begin{pmatrix}(|0\rangle+|1\rangle)|0\rangle - (|0\rangle+|1\rangle)|1\rangle + \\ (|0\rangle-|1\rangle)|0\rangle - (|0\rangle-|1\rangle)|1\rangle\end{pmatrix}$$

Distributing the second qubit across the superposition of the first yields:

$$= 1/2\sqrt{2}\,(|0\rangle|0\rangle + |1\rangle|0\rangle - |0\rangle|1\rangle - |1\rangle|1\rangle + |0\rangle|0\rangle - |1\rangle|0\rangle - |0\rangle|1\rangle + |1\rangle|1\rangle)$$

Collecting and canceling like terms, we get:

$$= 1/2\sqrt{2}\,(2|0\rangle|0\rangle - 2|0\rangle|1\rangle)$$

We now factor out the 2 and the first qubit.

$$= 1/\sqrt{2}\,|0\rangle(|0\rangle - |1\rangle)$$

Generalizing, we get the final result for this case.

$$= 1/\sqrt{2}\,|f(0)\oplus f(1)\rangle(|0\rangle - |1\rangle)$$

Case II: $f(0)=0 \quad f(1)=1$

$$1/2\sqrt{2}\begin{pmatrix}(|0\rangle+|1\rangle)|0\oplus f(0)\rangle - (|0\rangle+|1\rangle)|1\oplus f(0)\rangle + \\ (|0\rangle-|1\rangle)|0\oplus f(1)\rangle - (|0\rangle-|1\rangle)|1\oplus f(1)\rangle\end{pmatrix}$$

$$= 1/2\sqrt{2}\begin{pmatrix}(|0\rangle+|1\rangle)|0\rangle - (|0\rangle+|1\rangle)|1\rangle + \\ (|0\rangle-|1\rangle)|1\rangle - (|0\rangle-|1\rangle)|0\rangle\end{pmatrix}$$

$$= 1/2\sqrt{2}\,(|0\rangle|0\rangle + |1\rangle|0\rangle - |0\rangle|1\rangle - |1\rangle|1\rangle + |0\rangle|1\rangle - |1\rangle|1\rangle - |0\rangle|0\rangle + |1\rangle|0\rangle)$$

The reader can verify that collecting, canceling, and factoring gives:

$$= 1/\sqrt{2}\,|1\rangle(|0\rangle - |1\rangle)$$

This generalizes to case II's final result.

$$= 1/\sqrt{2}\,|f(0)\oplus f(1)\rangle(|0\rangle - |1\rangle)$$

Case III: $f(0) = 1 \quad f(1) = 0$

$$1/2\sqrt{2}\begin{pmatrix} (|0\rangle+|1\rangle)|0 \oplus f(0)\rangle - (|0\rangle+|1\rangle)|1 \oplus f(0)\rangle + \\ (|0\rangle-|1\rangle)|0 \oplus f(1)\rangle - (|0\rangle-|1\rangle)|1 \oplus f(1)\rangle \end{pmatrix}$$

$$= 1/2\sqrt{2}\begin{pmatrix} (|0\rangle+|1\rangle)|1\rangle - (|0\rangle+|1\rangle)|0\rangle + \\ (|0\rangle-|1\rangle)|0\rangle - (|0\rangle-|1\rangle)|1\rangle \end{pmatrix}$$

$$= 1/2\sqrt{2}\,(|0\rangle|1\rangle + |1\rangle|1\rangle - |0\rangle|0\rangle - |1\rangle|0\rangle + |0\rangle|0\rangle - |1\rangle|0\rangle - |0\rangle|1\rangle + |1\rangle|1\rangle)$$

$$= 1/\sqrt{2}\,|1\rangle(|1\rangle - |0\rangle)$$

$$= 1/\sqrt{2}\,|f(0) \oplus f(1)\rangle(|1\rangle - |0\rangle)$$

Case IV: $f(0) = 1 \quad f(1) = 1$

$$1/2\sqrt{2}\begin{pmatrix} (|0\rangle+|1\rangle)|0 \oplus f(0)\rangle - (|0\rangle+|1\rangle)|1 \oplus f(0)\rangle + \\ (|0\rangle-|1\rangle)|0 \oplus f(1)\rangle - (|0\rangle-|1\rangle)|1 \oplus f(1)\rangle \end{pmatrix}$$

$$= 1/2\sqrt{2}\begin{pmatrix} (|0\rangle+|1\rangle)|1\rangle - (|0\rangle+|1\rangle)|0\rangle + \\ (|0\rangle-|1\rangle)|1\rangle - (|0\rangle-|1\rangle)|0\rangle \end{pmatrix}$$

$$= 1/2\sqrt{2}\,(|0\rangle|1\rangle + |1\rangle|1\rangle - |0\rangle|0\rangle - |1\rangle|0\rangle + |0\rangle|1\rangle - |1\rangle|1\rangle - |0\rangle|0\rangle + |1\rangle|0\rangle)$$

$$= 1/\sqrt{2}\,|0\rangle(|1\rangle - |0\rangle)$$

$$= 1/\sqrt{2}\,|f(0) \oplus f(1)\rangle(|1\rangle - |0\rangle)$$

We compute the second qubit to be $|1\rangle - |0\rangle$ in cases I and II, and $|0\rangle - |1\rangle$ in cases III and IV. This extra multiplication by -1 is called a "global phase." A global phase is akin to rotating a cube in purely empty space: without a reference, it is merely a mathematical artifact and has no physical meaning.

Thus all the cases result in $1/\sqrt{2}\,|f(0) \oplus f(1)\rangle(|0\rangle - |1\rangle)$.

www.ingramcontent.com/pod-product-compliance
Lightning Source LLC
Chambersburg PA
CBHW081905170526
45167CB00007B/3162